$x(x+x^2)$?
x^2+x^3!
$\frac{-b \pm \sqrt{b^2-4ac}}{2a}$
$\frac{10}{6}$
44.9° North
93.2° West
If A ⊂ B and B ⊂ C
then A ⊂ C
7 PIZZAS
4n+2
OOOOH!
let's see... calculating... calculating...
WOW!
30×20 30×7 8×20 8×7
600 + 210 + 160 + 56 = 1026
No! Silence your nonsense!
"1000"
Guaranteed 100%
Almost certain 95% to 99.9%
Very likely 80% to 95%
Probable 60% to 80%
Maybe 40% to 60%
Possible 20% to 40%
Unlikely 5% to 20%
Almost impossible 0.1% to 5%
Impossible 0%
What is $\lim_{x\to 2} \frac{x^2-4}{x-2}$?
AROUND ≈ 3 X ACROSS
???
0 .1 .2 .3 .4 .5 .6 .7 .8 .9 1
What's 46 + 28?
27
$ax^2 = bx + c$
$ax^2 + bx = c$
$ax^2 + c = bx$
$ax^2 = bx$
$ax^2 = c$
$bx = c$
TRUTH
100°F
$1 \quad -1 \quad -(-1) \quad -(-(-1))$
1/3 sleep deprivation,
2/3 game theory.
$=\frac{1}{2}mv^2$

Math for English Majors

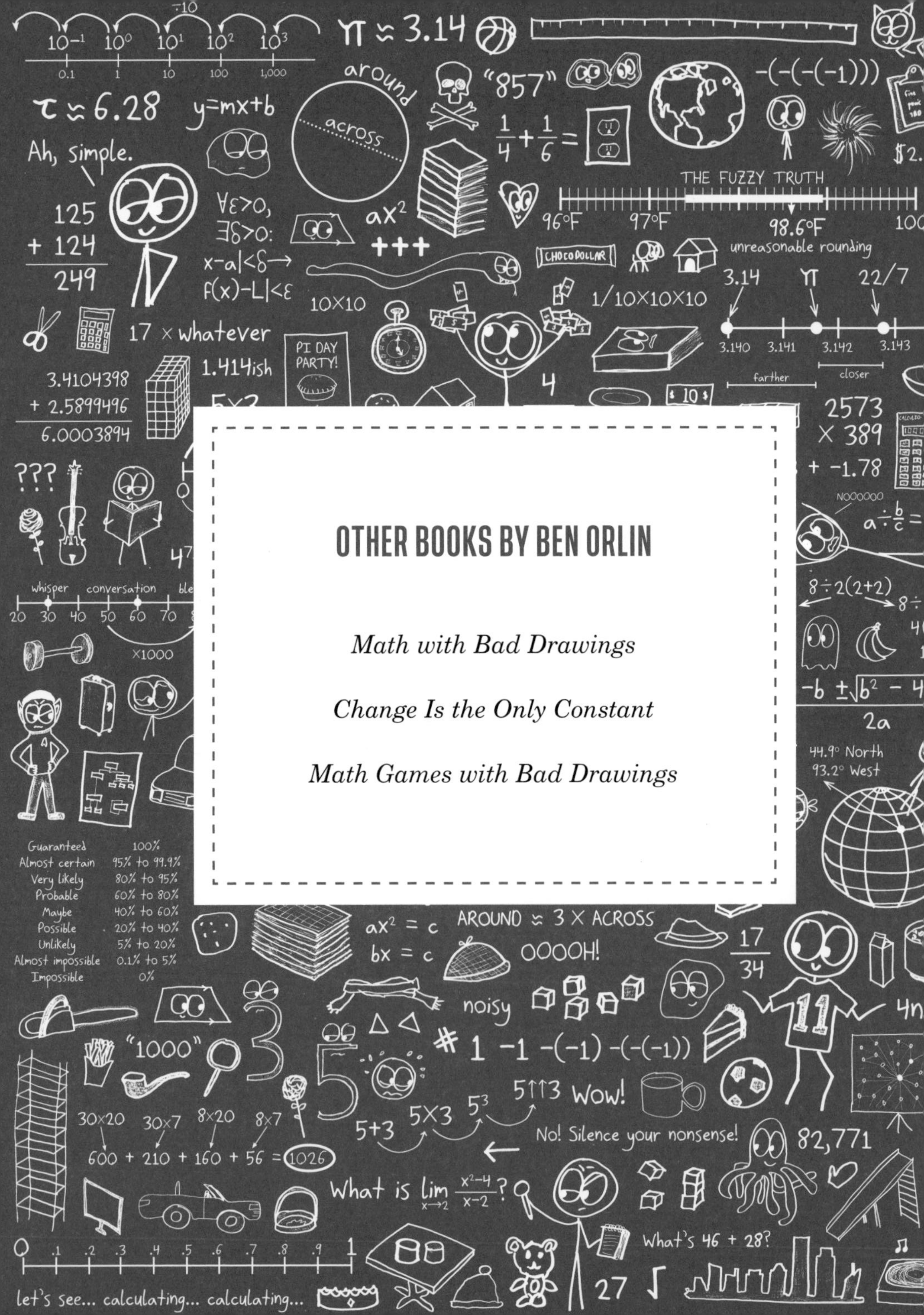

OTHER BOOKS BY BEN ORLIN

Math with Bad Drawings

Change Is the Only Constant

Math Games with Bad Drawings

Math ~for~ English Majors

A Human Take on
the Universal Language

BEN ORLIN

BLACK DOG
& LEVENTHAL
PUBLISHERS
NEW YORK

Black Dog & Leventhal Publishers
Hachette Book Group
1290 Avenue of the Americas
New York, NY 10104
www.blackdogandleventhal.com
BlackDogandLeventhal @BDLev

First Edition: August 2024

Black Dog & Leventhal Publishers is an imprint of Hachette Book Group, Inc. The Black Dog & Leventhal Publishers name and logo are trademarks of Hachette Book Group, Inc.

The Hachette Speakers Bureau provides a wide range of authors for speaking events. To find out more, go to hachettespeakersbureau.com or email HachetteSpeakers@hbgusa.com.

Black Dog & Leventhal books may be purchased in bulk for business, educational, or promotional use. For more information, please contact your local bookseller or the Hachette Book Group Special Markets Department at Special.Markets@hbgusa.com.

The publisher is not responsible for websites (or their content) that are not owned by the publisher.

Print book cover and interior design by Katie Benezra

Library of Congress Cataloging-in-Publication Data has been applied for.

ISBNs: 978-0-7624-9981-6 (hardcover), 978-0-7624-9983-0 (ebook)

Printed in China

1010

10 9 8 7 6 5 4 3 2 1

For Devyn,
whose facial expressions already say more
than any equations can

·10
10⁻¹ 10⁰ 10¹ 10² 10³
0.1 1 10 100 1,000
π ≈ 3.14
τ ≈ 6.28
y=mx+b
Ah, simple.
125
+ 124
249
around
across
ax²
+++
∀ε>0,
∃δ>0:
\|x−a\|<δ→
\|f(x)−L\|<ε
"857"
1/4 + 1/6 =
−(−(−(−(−1))))
THE FUZZY TRUTH
96°F
97°F
98.6°F
unreasonable rounding
CHOCODOLLAR
10×10
1/10×10×10
3.14 π 22/7
3.140 3.141 3.142 3.143
farther
closer
17 × whatever
1.414ish
3.4104398
+ 2.5899496
6.0003894
PI DAY PARTY!
3/14
5×3
4/7
10
2573
× 389
multiplication
0 5 10 15 20 25 30 35 40 45 50
repeated addition
division by 1/2
17
34
division by 2
3 + −1.78
???
5^{5^5}
repeated exponentiation
5↑↑3
tetration
λ
NOOOOOO
a ÷ b/c =
$4^{7/2}$ = ???
6
2
whisper conversation blender nightclub thunder
20 30 40 50 60 70 80 90 100 110 120
×1000
x(x+x²)?
x²+x³!
8 ÷ 2(2+2)
8 ÷ 2(4)
8 ÷ 8
1
x²=3x+10
K=½mv²
If A<B and B<C
then A<C
$x = \frac{-b \pm \sqrt{b^2 - 4ac}}{2a}$
DONUTS
44.9° North
93.2° West
7 PIZZAS
10/6
ax² = bx + c
ax² + bx = c
ax² + c = bx
ax² = bx
ax² = c
bx = c
1/3 sleep deprivation,
2/3 game theory.
Guaranteed 100%
Almost certain 95% to 99.9%
Very likely 80% to 95%
Probable 60% to 80%
Maybe 40% to 60%
Possible 20% to 40%
Unlikely 5% to 20%
Almost impossible 0.1% to 5%
Impossible 0%
AROUND ≈ 3 × ACROSS
OOOOH!
17/34
noisy
11
"1000"
3
5
1 −1 −(−1) −(−(−1))
5↑↑3 WOW!
5+3 5×3 5³
30×20 30×7 8×20 8×7
600 + 210 + 160 + 56 = 1026
No! Silence your nonsense!
82,771
What is $\lim_{x\to 2} \frac{x^2-4}{x-2}$?
What's 46 + 28?
0 .1 .2 .3 .4 .5 .6 .7 .8 .9 1
let's see... calculating... calculating...
27

Their own utterances, which appear so simple
and transparent to themselves, are, in fact, enormously
complex and contain and conceal the vast
apparatus of a true language.

—OLIVER SACKS, *SEEING VOICES*

CONTENTS

NOUNS

The Things Called Numbers • 1

VERBS

The Actions of Arithmetic • 67

GRAMMAR

The Syntax of Algebra • 129

PHRASE BOOK

A Local's Guide to Mathematical Vocabulary • 199

10⁻¹ 10⁰ 10¹ 10² 10³
0.1 1 10 100 1,000
π ≈ 3.14
"857"
−(−(−(−(−1))))
around
across
τ ≈ 6.28
y=mx+b
Ah, simple.
125
+ 124
249
∀ε>0,
∃δ>0:
|x−a|<δ→
|f(x)−L|<ε
ax²
+++
1/4 + 1/6 =
THE FUZZY TRUTH
96°F
97°F
98.6°F
100°
unreasonable rounding
CHOCODOLLAR
1/10×10×10
10×10
17 × whatever
3.14 π 22/7
3.140 3.141 3.142 3.143
farther
closer
3.4104398
+ 2.5899496
6.0003894
1.414ish
PI DAY PARTY!
3/14
5×3
multiplication
4/7
$ 10 $
2573
× 389
division by 1/2
17 34
division by 2
3 + −1.78
???
0 5 10 15 20 25 30 35 40 45 50
repeated addition
5⁵⁵ → 5↑↑3
repeated exponentiation
tetration
λ
NOOOOOO
a ÷ b/c =
4⁷ᐟ² = ???
6
2
whisper conversation blender nightclub thunder
20 30 40 50 60 70 80 90 100 110 120
×1000
x(x+x²)?
x²+x³!
8÷2(2+2)
8÷2(4)
8÷8
1
x²=3x+10
K=½mv²
If A<B and B<C
then A<C
x = (−b ± √(b² − 4ac)) / 2a
DONUTS
7 PIZZAS
10/6
44.9° North
93.2° West
ax² = bx + c
ax² + bx = c
ax² + c = bx
ax² = bx
ax² = c
bx = c
1/3 sleep deprivation,
2/3 game theory.
Guaranteed 100%
Almost certain 95% to 99.9%
Very likely 80% to 95%
Probable 60% to 80%
Maybe 40% to 60%
Possible 20% to 40%
Unlikely 5% to 20%
Almost impossible 0.1% to 5%
Impossible 0%
AROUND ≈ 3 × ACROSS
OOOOH!
17/34
noisy
11
4n+
"1000"
3
5
1 −1 −(−1) −(−(−1))
5↑↑3 Wow!
5+3 5×3 5³
30×20 30×7 8×20 8×7
600 + 210 + 160 + 56 = 1026
No! Silence your nonsense!
82,771
What is lim_{x→2} (x²−4)/(x−2)?
What's 46 + 28?
0 .1 .2 .3 .4 .5 .6 .7 .8 .9 1
let's see... calculating... calculating...
27

INTRODUCTION

I once asked an auditorium of undergraduates to recall their earliest memories of math. One of them shared a scene so peculiar—and yet so universal—that it burrowed deep into my unconscious, to the point where it has begun to feel like a memory of my own.

At age five, this student was assigned worksheets of addition problems. Trouble was, she didn't know how to read the funny symbols on the pages, the 2's and +'s and such. No one had ever taught her. Too intimidated to ask, she found a work-around, memorizing each sum not as a fact about numbers, but as an arbitrary rule about shapes. For example, $8 + 1 = 9$ was not the statement that 9 is one more than 8, but a coded set of instructions: if you are shown two stacked circles (8), followed by a cross (+), a vertical line (1), and a pair of horizontal lines (=), then you must fill in the blank space provided by writing a circle with a downward-curling tail (9). With painstaking diligence, she taught herself dozens of rules like this, each as baroque and pointless as the last. It was mathematics by way of Kafka.

Few people learn $8 + 1 = 9$ this way. But sooner or later, almost every math student suffers a similar sense of confusion and resorts to similarly desperate work-arounds. Whether in preschool, middle school, or grad school, befuddlement eventually descends, and mathematics becomes, in the words of mathematician David Hilbert, a game of "meaningless marks on paper."

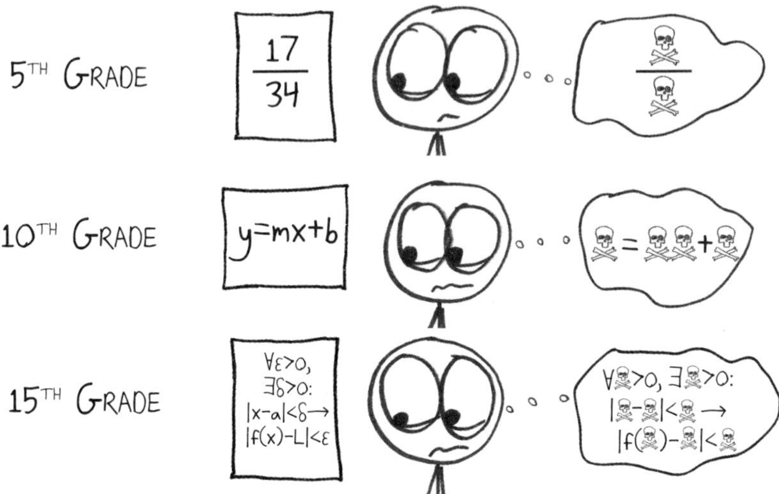

We've all been there. There's a symbol you don't recognize, a step you can't follow, so you ask what the mess of marks is supposed to mean. The reply is a stream of gibberish. So you ask what *that* means. The reply is a torrent of nonsense. This continues, frustration building on all sides, until at last you nod, smile, and say, "Oh, yes. *Thank* you. That clears everything up." Then, with all sense of meaning thwarted, you set about the grueling work of memorizing which shapes to write in what order.

Math, we like to say, is a language. (A "universal language," even.) But if a language brings people together, then why does math make us feel so alone?

I am a professional apologist for mathematics. I use "apologist" both in the classical sense (an advocate; a proponent; an expounder of a worldview) and in the modern sense (someone doing public relations for a widely despised client). What brought me to this career—the reason I became a math teacher—was the vague and overblown conviction that mathematics needed my help. Something was wrong, everyone seemed to say, terribly and perhaps fatally wrong, with the way that we teach math.

What, exactly, was wrong? Ah, that's where the consensus dissolved. I've spent the last 15 years trying to puzzle it out.

One common complaint is a lack of "real-world applications." Math is too abstract, too obscure, too far up its own ivory tower. As the timeless refrain goes: "When am I going to use this?" Many textbook authors take this grievance to heart. For example, they'll turn a question about a quadratic (*boring!*) into one about a company whose revenue is, in defiance of all rhyme and reason, a quadratic (*so real, so practical!*). Other educators reject the premise of the "real-world" complaint. No one asks when they'll "use" music or literature, do they? So why not follow the wisdom of Albert Einstein, and embrace math as "the poetry of logical ideas"?

No matter how we respond to the "real-world" concern, I suspect we're taking it too literally. When students ask for usefulness, they don't mean a sense of *practicality*. They mean a sense of *purpose*. "When will I use this?" means something like "What are we doing here?" or "Why does this stuff matter?" or "What does it all mean?"

They're not saying: "Describe the distant date when these exercises will benefit my bank account." Nor: "Explain the unlikely way in which these exercises might benefit my soul." The question is more like: "Tell me, here and now, what exactly *are* these exercises?"

Math is more than a collection of ideas. It's a specialized way of talking about those ideas. What the students are asking for, without realizing it, is help learning humanity's strangest language.

So what does it mean to say that math is a language?

Math begins with numbers. Though numbers and words differ in a few notable ways, both are systems for labeling the world. Numbers, like words, let us reduce a complex experience (say, a lakeside stroll) to something far simpler. In the case of words, a description ("There were lots of expensive dogs"); in the case of numbers, a quantity ("3 miles").

After numbers come calculations. Calculations generate new numbers from old numbers, which is to say, new knowledge from old knowledge. For example, if our 3-mile lake is a rough circle, then I can calculate the distance across the lake to be approximately 1 mile.

So far, so good. But then comes algebra.

Algebra, like literature or philosophy, takes a step back from the everyday world. We leave behind particular numbers (177) and particular calculations ($177 \div 3$) to study the nature of calculation itself. Algebra opens up new possibilities: streamlining computations, rearranging steps, comparing approaches, and so on. This requires a rich grammar, with a distinctive system of noun phrases and a small stable of workhorse verbs. Most notably, specific numbers like 3 give way to abstract placeholders like x. This leap of faith from the concrete 3 to the general x marks the dawn of a whole new language—and for many people, the dusk of understanding.

This little book has a lofty aim: to teach you the language of math. We'll build from the abstract nouns of number to the active verbs of calculation to the nuanced grammar of algebra. A few cartoon-illustrated pages cannot teach the entire language, of course, but I hope they can give you a running start.

What I am proposing is a little out of the ordinary. When mathematicians write for a general audience, we tend to celebrate the subject's ideas and applications, not the language in which they are expressed. Often we abandon the language altogether, translating the equations (as best we can) into English prose.

This book takes a more brambly and less trammeled path. It is not literature in translation, but an attempt to animate the beautiful and austere language that makes such literature possible.

A classic riddle asks whether math was discovered or invented. Is math out there in the fabric of nature? Or is it a tool we created for the sake of examining nature? Is mathematics the atom or the microscope?

My reply, of course, is both. Mathematics is an invention wrapped around a discovery; it is a house built around a tree. The house is a language so cleverly crafted that it feels like a work of nature. The tree is a discovery so magical in its architecture that it feels like a work of design. Mathematics is the atom *and* the microscope, unified so seamlessly it can be hard to tell where discovery ends and invention begins.

This intertwining of invention and discovery, of language and idea, is one reason math can be so hard to learn. To grasp the ideas, you must first learn the language, but the language makes no sense except as an expression of the ideas.

It was never my plan to make a career as a math apologist. If I have been guided toward math, then I was not like a Greek hero whom the gods usher to his fate but more like a confused tourist whom the locals usher out of traffic.

Still, here in the tree house, watching the light through the leaves, I can't help wishing that everyone could stand right where I'm standing. I hope this little book can bring you here.

÷10
10⁻¹ 10⁰ 10¹ 10² 10³
0.1 1 10 100 1,000
π ≈ 3.14
τ ≈ 6.28
y=mx+b
Ah, simple.
125
+ 124
249
∀ε>0,
∃δ>0:
x−a|<δ→
f(x)−L|<ε
around
across
"857"
1/4 + 1/6 =
ax²
+++
-(-(-(-(-1))))
THE FUZZY TRUTH
96°F
97°F
98.6°F
100
unreasonable rounding
CHOCODOLLAR
10×10
1/10×10×10
3.14
π
22/7
3.140 3.141 3.142 3.143
farther
closer
17 × whatever
1.414ish
3.4104398
+ 2.5899496
6.0003894
PI DAY PARTY!
3/14
5×3
multiplication
4/7
10
division by 1/2
17
34
division by 2
2573
× 389
3 + −1.78
???
0 5 10 15 20 25 30 35 40 45 50
repeated addition
5⁵⁵
repeated exponentiation
5↑↑3
tetration
NOOOOOO
a÷b/c =
4⁷ᐟ² = ???
6
2
whisper conversation blender nightclub thunder
20 30 40 50 60 70 80 90 100 110 120
×1000
x(x+x²)?
x²+x³!
8÷2(2+2)
8÷2(4)
8÷8
1
x²=3x+10
K=½mv²
If A<B and B<C
then A<C
x = −b ± √b² − 4ac
2a
DONUTS
44.9° North
93.2° West
7 PIZZAS
10/6
ax² = bx + c
ax² + bx = c
ax² + c = bx
ax² = bx
ax² = c
bx = c
1/3 sleep deprivation,
2/3 game theory.
Guaranteed 100%
Almost certain 95% to 99.9%
Very likely 80% to 95%
Probable 60% to 80%
Maybe 40% to 60%
Possible 20% to 40%
Unlikely 5% to 20%
Almost impossible 0.1% to 5%
Impossible 0%
AROUND ≈ 3 × ACROSS
OOOOH!
17/34
noisy
11
"1000"
3
5
1 −1 −(−1) −(−(−1))
5↑↑3 WOW!
30×20 30×7 8×20 8×7
600 + 210 + 160 + 56 = 1026
5+3
5×3
5³
No! Silence your nonsense!
82,771
What is lim x→2 (x²−4)/(x−2)?
What's 46 + 28?
0 .1 .2 .3 .4 .5 .6 .7 .8 .9 1
27
let's see... calculating... calculating...

NOUNS

The Things Called Numbers

A "noun" is typically defined as "a word for a person, place, or thing." When I was a kid, this definition always troubled me. It seemed obvious to me that people and places are also *things*, so why the redundancy? Why not just define a "noun" as "a word for a thing" and be done with it? Looking back, I can see that this childhood flight of pedantry exemplified one of the peculiar principles of mathematical language: everything is a thing.

Case in point: *numbers*. They are the oldest and most familiar things in math, but they aren't really things at all. Travel the world, and you may cross seven seas, taste seven pizzas, or battle seven ninjas, but you will never encounter any such *thing* as "seven."

There is no "seven," only seven *of* something. Saying "seven marbles" is like saying "blue marbles"—a property, a descriptor. Not a noun, but an adjective.

Or so a reasonable person would say. But a mathematician is hardly a reasonable person. More like a feral philosopher or a logician gone rogue.

In the same way the adjective "beautiful" gives rise to a noun "beauty," the adjective "seven" gives rise to a noun also (confusingly) called "seven" and defined by the elusive quality of seven-ness, the trait that all groups of seven items share in common.

Thus, a number is a noun born from an adjective. It is an intangible property so compelling that we study it for its own sake, as if it were a thing itself. As we'll later see, numbers are not the only nouns in math, but they are the most essential, and will thus occupy the first section of this book.

The novelist Karen Olsson describes math as "a cloud land of tantalizing abstract structures, curves and surfaces and fields and vector spaces, accessible only to those who learn the elaborate cloud language, a vehicle for truths that cannot be expressed in any other tongue."

As befits a cloud language, we begin our study inside a cloud. Join me for a stroll in the forests of Snowdonia, on a mountainside wrapped in mists…

Counting

A few years ago, hiking on a hillside in Wales, I came across a plaque listing the Welsh-language numbers from 1 to 20. Being an avid fan of plaques, numbers, and the Welsh people, I dove right in.

One: *un*.

Two: *dau*.

Three: *tri*.

No surprises until I reached 16: *unarbymtheg*. This seemed to be made of *un* (1) and *pymtheg* (15). My English-speaking mind found it a peculiar and charming name, so I was pleased to see that 17 followed suit (*dauarbymtheg*, 2 and 15), and 19 as well (*pedwararbymtheg*, 4 and 15). I could guess by now what 18 would be: *triarbymtheg*, 3 and 15. Right?

No. The Welsh refused to indulge my pedestrian logic. Eighteen was *deunaw*: literally, two 9's. I stood in the mists of Snowdonia, my heart swelling with admiration for the people of Wales and for the number they had so beautifully named.

To name something is to set it apart, to give it an identity. That's why we name babies, songs, cities, pets, and group chats—but we don't name, say, individual paperclips. I'm keen to distinguish my baby from yours; I'm less worried about my office supplies.

Without a name, a number is not really a number. It's something more like a paperclip, indistinguishable from all the others. How easily can you tell •••••••••••••••••• from ••••••••••••••••• or •••••••••••••••••••? Mathematics can only begin when each number gains a name—and thus, an identity. In the book of Genesis, Adam names the creatures of the Earth, from aardvark to zebra. In the 18th century, botanist Carl Linnaeus did the same, from *Orycteropus afer* to *Equus quagga*. What Adam and Carl did for life-forms, we must do for quantities. This process of naming the numbers, one by one, is known as counting.

In English, the number •••••••••••••••••• is named "eighteen": literally, "eight and ten." That's an accurate description, but •••••••••••••••••• is also three 6's, or a dozen and a half, or nine pairs. Why say "eighteen" when there are more pleasing alternatives? Why the clumsy, lopsided "eight and ten," and not the vivid symmetry of "deunaw"?

The question strikes at a deeper one. What exactly do we want from a counting system?

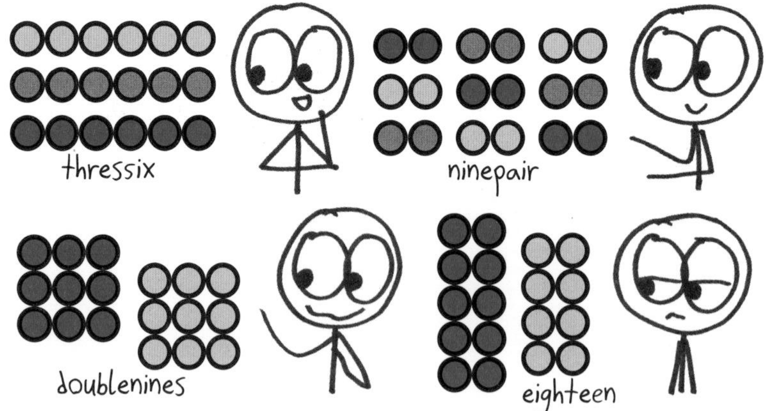

The short story "Funes the Memorious," by Jorge Luis Borges, tells of a boy who is thrown from a horse and knocked unconscious. Funes wakes with a curse and a gift: His body is paralyzed, but his mind is full of pictures. Anything he sees, he sees forever, in perfect detail. And so, as he lies in bed, Funes invents his own system of enumeration. To each number, he assigns a specific image: "sulfur," "the reins," "Napoleon." In his way of counting, every name is magnificent and distinct.

But as the narrator tries in vain to explain to Funes, such a math is no math at all.

Our naming system, known as "base ten," is based on breaking everything into groups of tens. A hundred is ten tens; a thousand is ten tens of tens; a million is ten tens of tens of tens of tens of tens. With all numbers built from the same standard parts, it is easy to compare and calculate with numbers: for example, it's easy to see that 125 is one more than 124, and easy to add the two (100 + 100, and 20 + 20, and 4 + 5) to arrive at a total of 249.

Not so with Funes's numerals. How is one to determine that "Máximo Pérez" is followed by "the train," or that their sum is "a cracked red brick"? As the tale's narrator puts it: "This rhapsody of unconnected terms was precisely the contrary of a system of enumeration."

This is why we pass up the poetic "deunaw" in favor of the prosaic "eighteen." With infinite numbers to name, we need an organized system, and our system runs on tens.

There is nothing inherently special about ten. We just happen to descend from ten-fingered apes. If we organized our system by eights, as octopuses or spiders might, then we'd call "eighteen" 22: two 8's plus two leftovers.

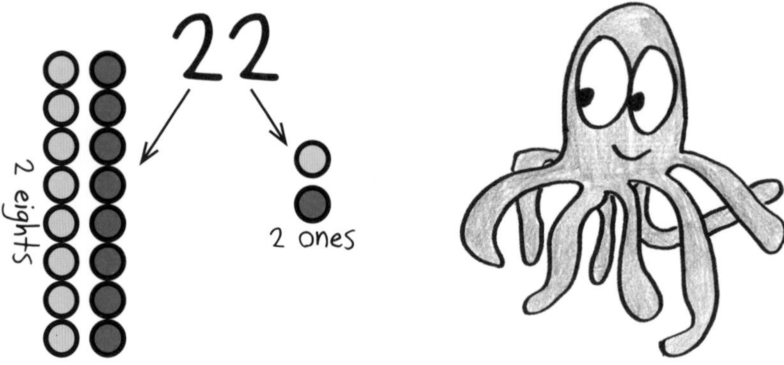

Or if we grouped by sevens, we'd call it 24: two 7's plus four leftovers.

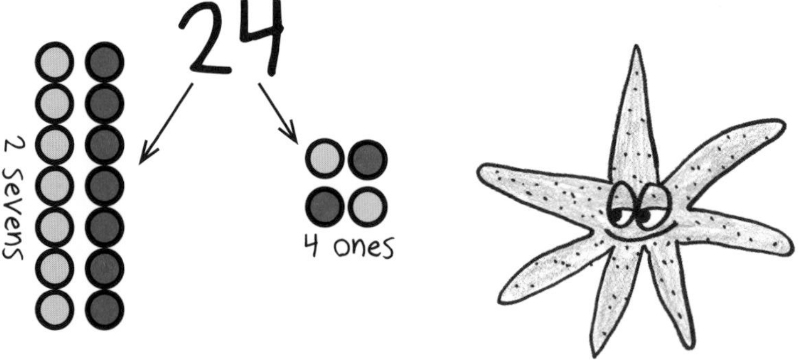

Or if we grouped by nines, we'd call it 20: two 9's plus zero leftovers.

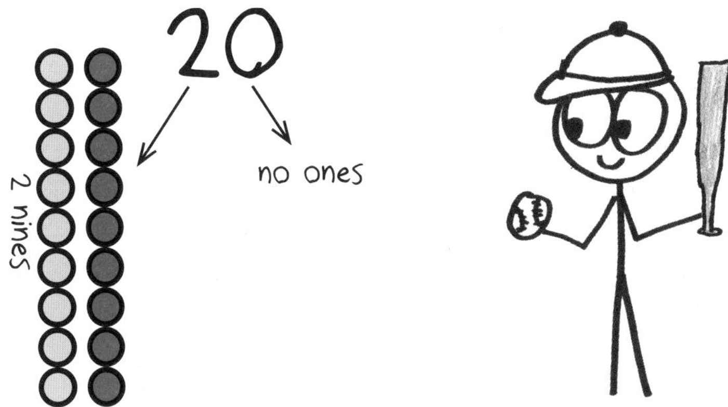

That's deunaw, isn't it? Yes, but at a price. To rename 18 as 20, we'd have to abandon our language of tens (10), hundreds (10 × 10), and thousands (10 × 10 × 10). In their place, we'd break numbers into nines (9), 81's (9 × 9), and 729's (9 × 9 × 9). The move would transform our whole naming system.

Seven hundred (seven groups of 10 × 10) would no longer have a nice round name. Its title would be the dull 857: eight groups of 9 × 9, five groups of 9, and seven leftovers.

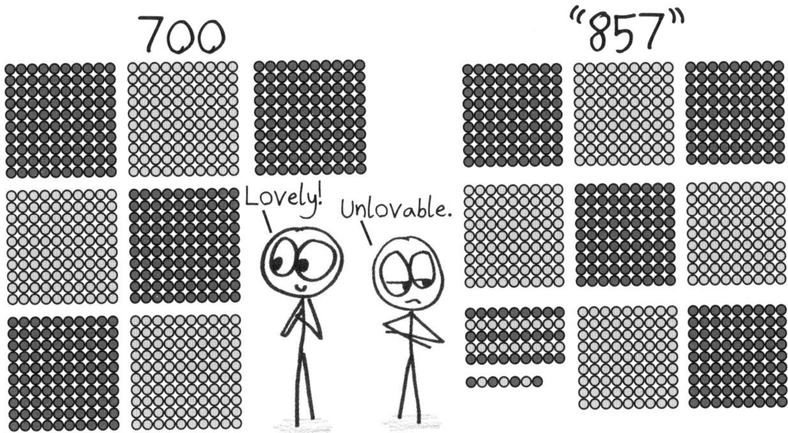

Meanwhile, the unremarkable 729 (seven groups of 10 × 10, two groups of 10, and nine leftovers) would become the beautifully round 1000: one perfect group of 9 × 9 × 9.

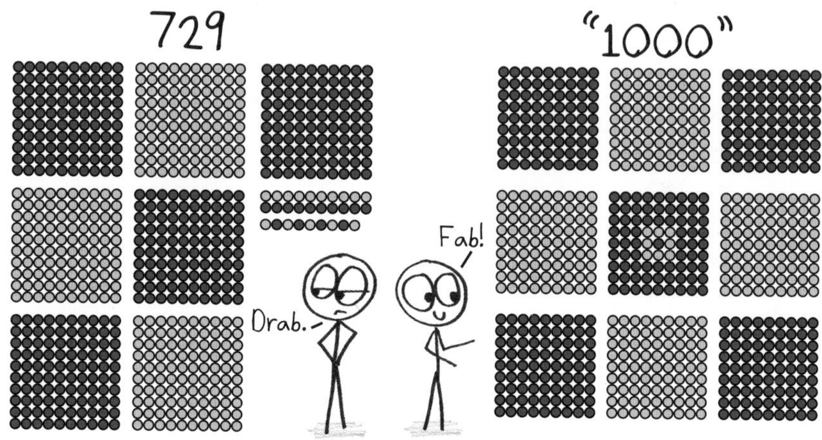

The numbers themselves have not changed—only their names. Yet names shape our world. In the world of deunaw, high school classmates would embrace at 27th reunions (now called *30th reunions*). Towns would throw lavish parades to commemorate 81 years (the base-nine *century*) since their founding. Drivers would pull over to snap photos when their odometers hit 59,049 (the digits reading 100,000).

Numbers mean so much to us: 10th birthdays, 50th anniversaries, 200th memorials. But is it the numbers we cherish, or only the names we've given them?

In Ursula Le Guin's Earthsea novels, there is a mystical language of true names. A thing and its name are somehow one and the same, so that knowing a person's true name grants you mortal power over them. Sometimes I think of math this way: I write the name 18, followed by the name 61, and then a modest bit of magic reveals that their sum is 79. Like an Earthsea sorcerer, I am able "to summon a thing that is not there at all, to call it by speaking its true name."

Alas, Earthsea is fantasy. In the real world, we must choose between half-truths. On the one hand, a language of orderly and systematic names; on the other, a language of vivid and memorable ones. On the one hand, the dull asymmetry of "eighteen"; and on the other, the Welsh perfection of "deunaw."

Measurement

Digging through our luggage on a Thanksgiving trip to Boston, my then three-year-old daughter found the thermometer I'd packed. "Oh!" she said. "I know how to do this." I watched as she tucked the instrument under her arm, waited a moment, and pulled it out for inspection. "Thirty point pounds," she announced. "I'm getting taller!"

Admittedly, her laboratory methods need polish. Still, at her tender and ungovernable age, she had already hit upon the foundation of mathematical language: quantification.

To *quantify* is to translate the world into numbers. We start with reality: the enigmatic and irreducible fabric of our existence. Then, being human, we assign it a numerical score. We reduce long things to a number called *length*, heavy things to a number called *weight*, and intelligent things to a number called *exam scores*. Such quantification knows no bounds (and no decency). Every week, it seems, a new and hitherto-unquantified part of life—something like nostalgia, or grief, or noodle soup—falls into the clutches of an enterprising misanthrope and is promptly turned into a number.

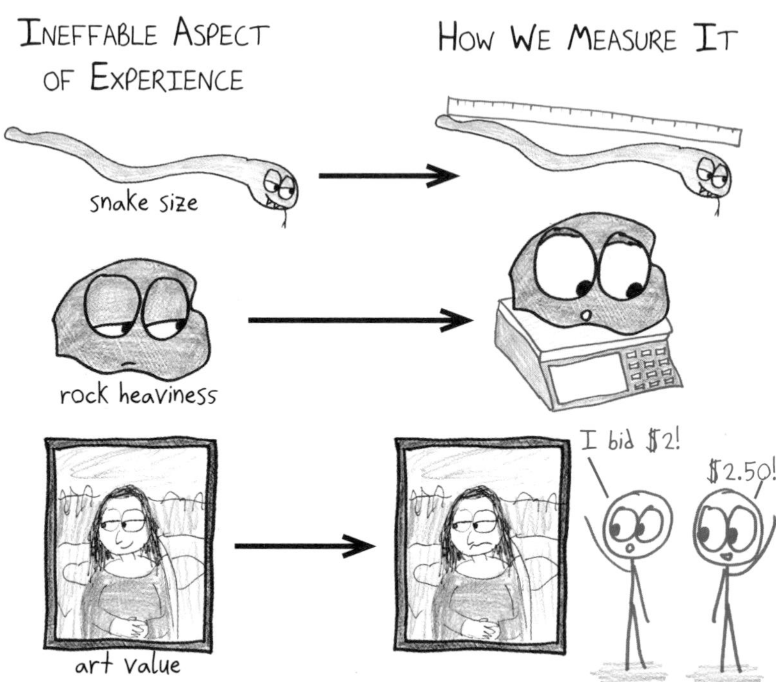

Another word for quantification is *measurement*. And measurement, as we know, requires tools. To measure time, you need a stopwatch; to measure public opinion, a survey or poll; to measure temperature, a thermometer; and to measure height or weight, apparently also a thermometer. Even the simplest acts of measurement—say, counting bath toys or remembering one's age—require the device of a pointing finger or a wall calendar.

Measurement is never perfectly accurate. For years, I told people I was 5'9", until one day I looked at my driver's license to learn that I am legally 5'8". It's not (just) self-delusion; it's that my height falls between the two values, and, even worse, it changes depending on how you measure it. Shoes add a centimeter, socks add a millimeter, and you must consider even seemingly irrelevant factors such as the time of day. (Gravity compresses one's spine ever so slightly, so that we are taller in the morning and shorter in the evening.) In any case, the lines on a tape measure are ⅓ of a millimeter thick, so you can never be more accurate than that.

No act of measurement is immune from error. The world's most accurate clock loses a nanosecond every year or two. Even counting itself is fallible. Give a typical person a jar of jelly beans, and their inevitable slips of attention will throw their count off by 1% or so.

The results of a measurement look squeaky-clean, yet they emerge from an inherently messy process. In that sense, measurement is like nothing so much as money laundering.

I find it a bit surprising, given all this, that mathematicians don't tend to dwell on measurement. Indeed, in explaining the nature of numbers, they rarely invoke measurement at all.

Case in point: *negative* numbers. You can't count –3 dogs, walk –3 miles, or sleep for –3 hours. In fact, no measurement process will ever give –3 as a result (unless we've rigged it by writing "–3" on the thermometer, even though the mercury has risen a positive distance).

If numbers come from measurement, then where does −3 come from?

The same goes for *irrational* numbers, such as $\sqrt{2}$ and π. To measure an irrational length, you'd need a tape measure of infinite precision. Impossible. But if no measurement ever yields an irrational number, then in what sense is an "irrational number" a number at all?

Then there's the case of *imaginary* numbers (such as *i*, the square root of −1). The name "imaginary" originated as a slur, from a mathematician who refused to accept their existence. These numbers are found not *on* the number line, but *above* and *below* it. Weird. Definitely not measurements. Yet they're numbers.

Aren't they?

Absolutely. Negative, irrational, and imaginary numbers emerge quite naturally—not from measurement itself, but from patterns and calculations among measurements. Subtract 8 from 5, and—bam—the result is *negative*. Compute the diagonal length across a square, and—bang—the distance is *irrational*. Solve a simple equation like $x^2 = -1$, and—boom—the solution is *imaginary*. The language of numbers may begin with measurement, but the numbers quickly take on a life of their own.

One day, my daughter will learn about the whole bizarre zoology of nonmeasurement numbers. Somewhat sooner, she will need to learn that underarm temperatures are not very accurate, especially not for determining heights and weights. But for now, I'm happy that she has grasped an elemental truth: mathematical language begins when you thrust a thermometer under the arm of reality and emerge with a number to announce.

Negative Numbers

In high school, I had a classmate infamous for derailing lessons with wild tangents. "Hey," he once said, "I add something to this class. It may be a negative number, but it's still an addition operator in my book."

I always loved that line. It captures the spirit of negative numbers: The presence of an absence. A lot of a lack.

Negative numbers are a linguistic trick, a way of unifying opposites. "A hill can't be a valley," Alice tells the Red Queen. "That would be nonsense." Yet with negative numbers, a hill *is* a valley—or rather, a valley is a negative hill. Instead of "300 feet below sea level" or "14,000 feet above sea level," we call the altitudes –300 and +14,000 (sometimes omitting the +). Similarly, instead of "eight minutes after the rocket launch" or "15 minutes before the launch," we call the times +8:00 and –15:00. Certain prepositions ("after," "above," "forward," "up") translate into +, while their opposites ("before," "below," "backward," "down") convert into –.

Negative and positive are mirror images that join to form the continuum known as *the number line*. Since the number line gained currency in the 1600s, its novelty has faded, but not its power.

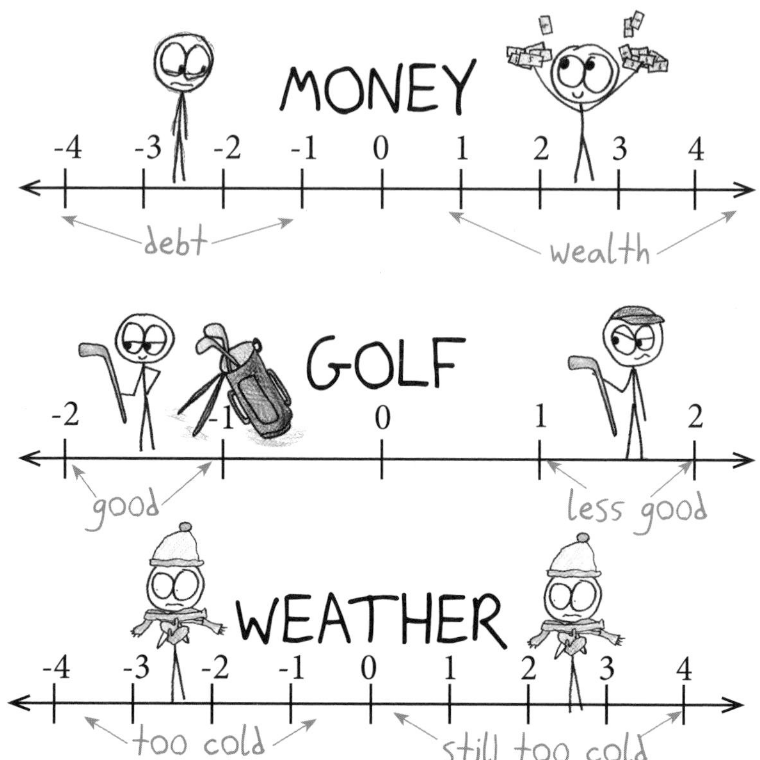

Today, we call the counting numbers (1, 2, 3, and so forth) *natural* numbers. Throw in zero and the naturals' opposites (–1, –2, –3, and so forth) and we call this entire group *integers*. Simple enough. So why did centuries of mathematicians refuse to accept negatives as genuine numbers? Why did Michael Stifel call them "absurd" and "fictitious"? Why did Bhaskara note that "people do not approve" of them? Why did Francis Maseres deem them "mere nonsense or unintelligible jargon"?

It all boils down to a simple question: How could a person with only $2 to their name manage to spend $3?

Anyone with a credit card can tell you the answer: all too easily.

Negative dollars are more commonly known as "debt." To form a mental picture of negative numbers, imagine that the *greenback* $1 bill (+) has a devilish counterpart: the *redback* negative $1 bill (–), signifying a dollar owed. Mixing the two kinds of bills allows us to visualize the same amount of money in various ways.

Such visuals allow us to make sense of arithmetic with negative numbers. For example, adding a positive number to your wallet—that is, acquiring greenbacks—is always a gain. It improves your bottom line.

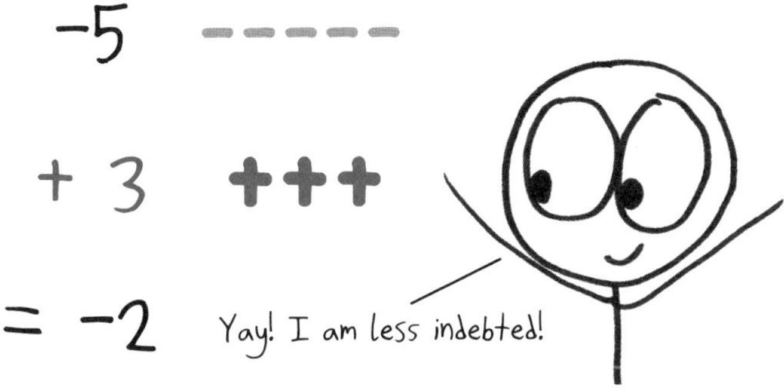

Meanwhile, subtracting a positive number—that is, giving up greenbacks—is always a loss. It leaves you worse off than before. In fact, if you started with a mix of greenbacks and redbacks (say, seven green and one red) then losing enough greenbacks may land you in debt.

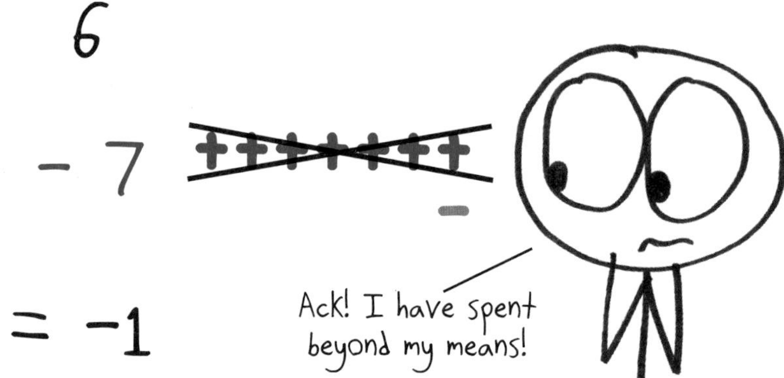

6
− 7
= −1
Ack! I have spent beyond my means!

Now, what about adding a negative? That is, acquiring red bills? It's bad news. A gain that impoverishes you.

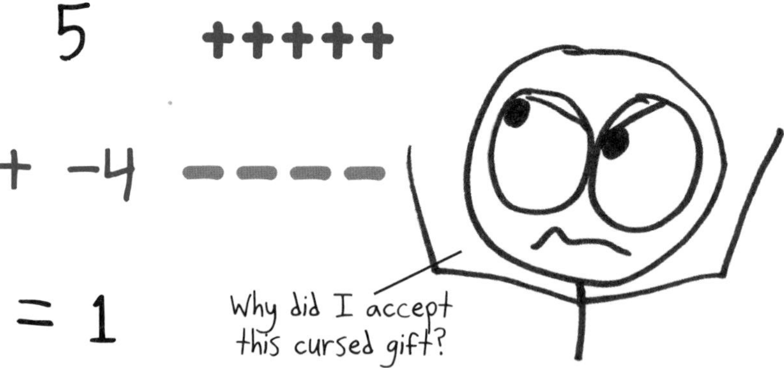

5
+ −4
= 1
Why did I accept this cursed gift?

On the other hand, to subtract a negative—that is, to offload red bills—is a gain. It's a loss that leaves you richer. If a person has some assets (say, $7) and some debts (say, $3), then taking away the debt causes their net worth (already positive) to rise further.

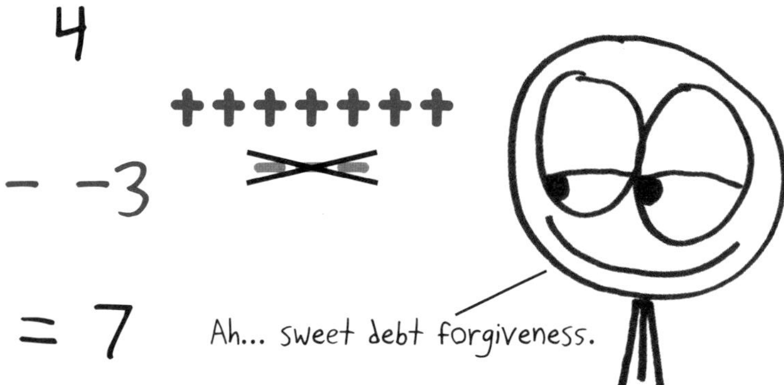

In this system, two symbols (+ and –) stand for four ideas: positive numbers (+), negative numbers (–), the operation of addition (+), and the operation of subtraction (–). When symbols work multiple jobs, mathematicians call them "overloaded." This kind of overloading drives some teachers to desperation and wrath; they'll storm like a thundercloud over any student who'd conflate "negative 7" (the number) with "minus 7" (the operation).

I admire the passion but don't share the philosophy. Not only are –7 (the negative) and –7 (the subtraction) easy to confuse, but they're *meant* to be confused. Ignoring the distinction reveals a lovely pattern, just as you see a hidden Magic Eye image only by letting your vision go fuzzy:

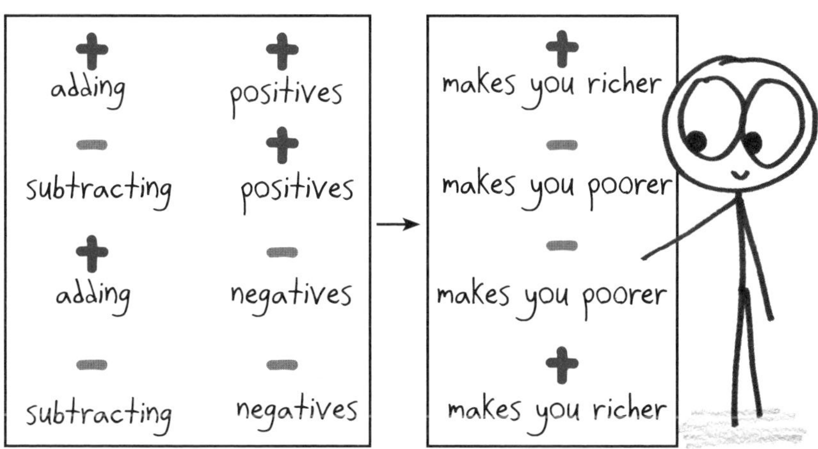

This table explains the bizarre yet common claim that "two negatives make a positive." Taken as a general rule, it's nonsense. Borrowing $20, then borrowing $30 more does not leave you $50 richer. The poet W. H. Auden recalled an old jingle from his school days: "Minus times minus equals plus. / The reason for this we need not discuss."

Well, the reason—which we *do* need discuss—is that an opposite's opposite is the thing itself. The opposite of day is night. The opposite of night is day. Therefore, the opposite of "the opposite of day" is day.

When I say "don't not jump," I'm asking you to jump. By the same logic, a negative times a negative is the opposite of an opposite.

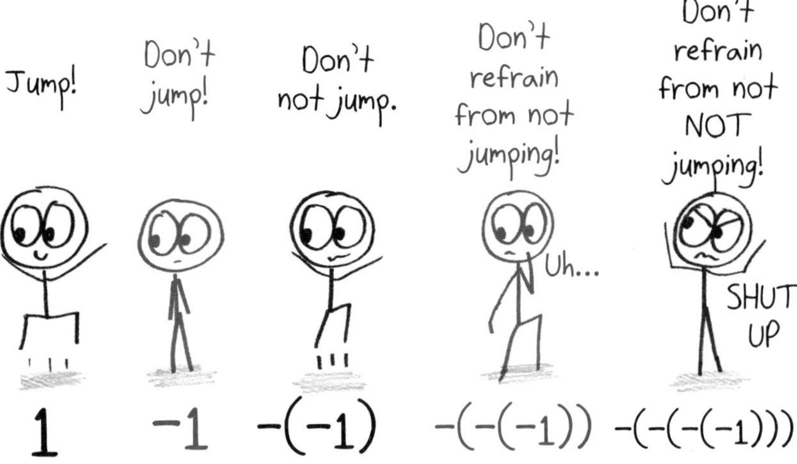

When I run this logic by students, they react as if I've shown them a nifty card trick. That is to say, they're not wholly convinced; clearly, it seems, some sleight of hand was involved. I'm then forced to point out that negatives are *entirely* sleight of hand—that, in fact, it's perfectly reasonable to reject negatives altogether.

Just ask al-Khwārizmī, the 9th-century mathematician often hailed as the father of algebra. He demonstrated how to solve quadratic equations, but if you showed him the form in today's textbooks, $ax^2 + bx + c = 0$, he'd spit medieval tea in your face. "What blasphemy is this?" he would say. "You're saying that three numbers added up equal zero? Some apples plus more apples plus more apples yields a total of…no apples?"

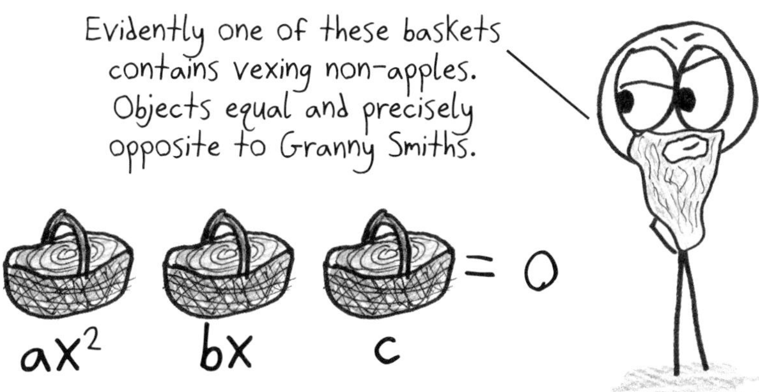

Evidently one of these baskets contains vexing non-apples. Objects equal and precisely opposite to Granny Smiths.
= 0
ax2
bx
c

It makes more sense for one basket to equal another. Or perhaps two baskets put together could equal a third. That's how al-Khwārizmī saw things—and that's why he solved not a single quadratic equation, but six of them, each carefully arranged to avoid negative quantities:

You know, you could save a lot of trouble if you just…

I confess that negatives tax the imagination. But if imagination won't pay, convenience suffers six times over, as one headache splits into half a dozen.

Positive numbers are great for counting and measuring. But negatives have a different job: to create a unified, harmonious system. Two concepts ("above sea level" and "below sea level") become one ("altitude"). Two operations ("addition" and "subtraction") become one (in modern algebra, all subtraction is reimagined as "adding the opposite"). And six formulas (al-Khwārizmī's nitty-gritty cases) become one (the sleek modern generalization).

In this way, negatives improve math by streamlining it. They add by subtraction. They're like the old joke about a party guest so negative that his arrival prompts others to ask, "Wait, who just left?" Perhaps it's a strange kind of betterment, but it's still an addition operator in my book.

Fractions

Over drinks at a holiday party, a professor and I fell into lamenting the original sin of math education, the failure that breeds a thousand failures: fractions. For too many students, fractions are a doubt that slowly grows, a mist that never lifts.

"What they need to understand," this professor insisted, "what we need to teach them, is that first and foremost, fractions are equivalence classes."

I disguised my laughter as a cough. *Sure, pal. Brilliant plan. We should simply tell students that fractions are ordered pairs of integers* (a,b) *under an equivalence relation whereby* (a,b) = (c,d) *if and only if* ad = bc. *Why didn't I see it before?* I nodded along and updated my mental list of evidence that math professors are space aliens.

Years later, sitting down to find a common thread in my notes for this chapter, the truth hit me with a smack, as if I'd stepped on a rake.

What you need to understand is...well, it's that fractions are equivalence classes.

Fractions are a language of in-betweens. If you want to buy cakes in bulk—say, three, four, or 17 cakes—then whole numbers suffice. But to purchase cakes by the slice, you'll need fractions, which involve two numbers. The number below the fraction bar, the *denominator*, represents how many slices each cake has been split into. The number above the bar, the *numerator*, is the number of slices we're buying.

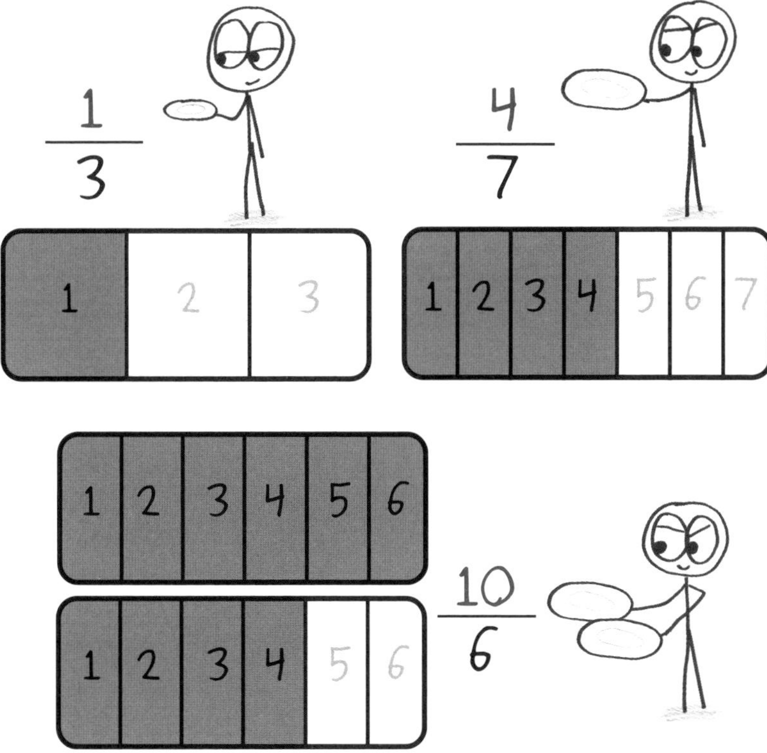

This language has a dizzying feature. Every fraction belongs to an infinite family of synonyms. Or, in mathematical terms, fractions form *equivalence classes*.

Say I desire half of a cake. I can slice it into two pieces and then eat one. Or I can slice it into four pieces and then eat two. Or six pieces and then eat three. Or eight hundred, and then eat four hundred. I need not worry about creating a crumbly mess; in math, fractions are abstractions. If I wish, I can cut 100 trillion invisibly thin pieces and then gorge myself on 50 trillion of them. Mathematically, the effect is the same as cutting two pieces and eating one.

This is why fractions wreak such havoc. How do you navigate a language of infinite synonyms?

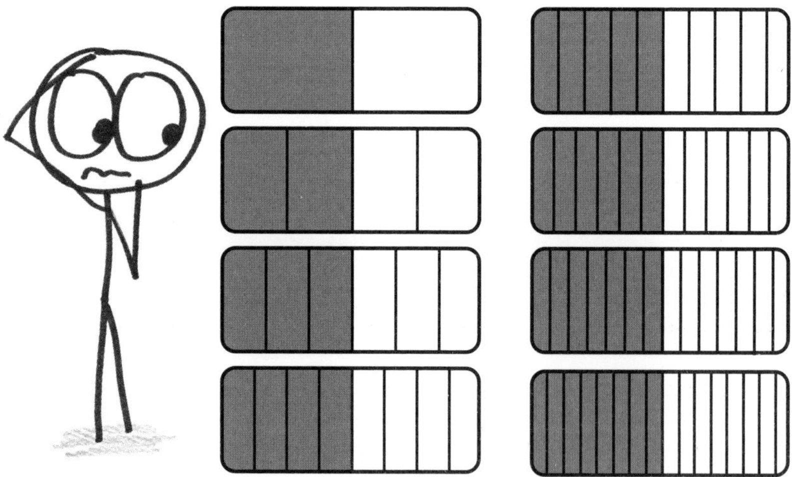

One way to escape the labyrinth of synonyms is never to cut more slices than necessary. Speak in terms of the lowest possible numbers. (We call these, fittingly enough, the "lowest terms.") Say $\frac{1}{5}$, not $\frac{2}{10}$. Say $\frac{3}{4}$, not $\frac{66,351}{88,468}$.

This simplification is often welcome, but sometimes self-defeating. In our effort to avoid the challenge of fractions, we end up missing out on their charm.

For example, when expressing $\frac{3}{4}$ of an hour, isn't it clearer to deploy the synonym $\frac{45}{60}$? Similarly, isn't $\frac{1}{4}$ of a century harder to parse than $\frac{25}{100}$? Why insist on making as few slices as possible when so many quantities come presliced?

Fraction synonyms are also essential for addition and subtraction. If I want $\frac{3}{4}$ of a cake, and you want $\frac{1}{6}$, then how much cake will feed us both? If we cut the cake into four slices, your slice is too big. Cutting it into six, my slices are too small. What to do?

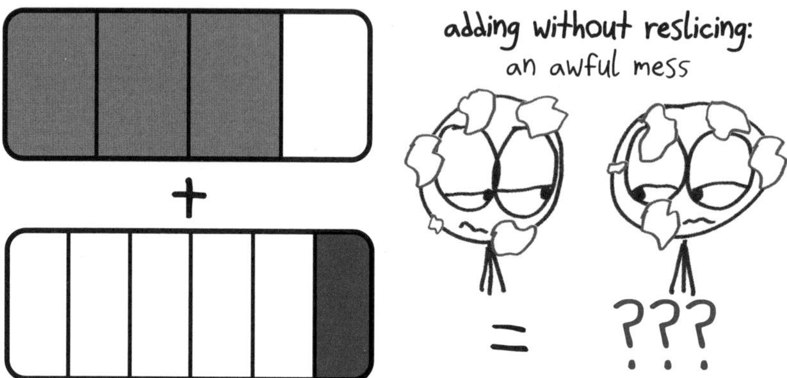

The solution is, we cut the cake so that it can be broken into four *or* six. Twelve pieces will do nicely. Now, my portion is $\frac{9}{12}$, and yours is $\frac{2}{12}$. Together, it's $\frac{11}{12}$. Synonyms save the day.

What about multiplying? Say I've got $4/5$ of a cake, and want to give you $2/3$ of what I have. Again, it helps to seek synonyms: that is, to slice the pie more finely. Take my current share (four pieces out of five); cut each piece into three thinner pieces (now I've got 12 out of 15); then give you two of every three (so you get eight out of 15). Mathematically, we just multiplied $4/5$ by $2/3$ to get $8/15$.

I should confess that no one actually multiplies fractions like this. There's an easier way: just multiply the tops and the bottoms. Do 2×4 and 3×5, then conclude that $\frac{2}{3} \times \frac{4}{5} = \frac{8}{15}$. Same great answer, way less thinking.

But this method's pleasure is the same as its peril: It conjures no mental picture. It's not a thought process so much as a thoughtless one. And thus, it breeds thoughtless errors.

For example, students who learn to "multiply across" often extrapolate to "adding across." They might proffer the absurdity that $\frac{1}{4} + \frac{1}{6} = \frac{2}{10}$, a heretical calculation, in which the whole is less than the sum of its parts. When students ask "Can I add across the fractions like this?" I am flooded with stress hormones, as if they'd asked "Can I put this fork in the wall socket?"

You never escape fractions. Show me a struggling high schooler, and I'll show you a secret discomfort with fractions, an unease that haunts them like a childhood memory.

Just ask the restaurant chain A&W. In the 1980s, their lavishly advertised new "third-pound" hamburger—selling for the same price as a McDonald's Quarter Pounder and, according to focus groups, tasting just as good—flopped. Folks weren't interested. "Why should we pay the same amount for a third of a pound of meat," they demanded, "as we do for a quarter pound?"

Whoever said "the customer is always right" must not have had clients claiming $\frac{1}{3}$ is smaller than $\frac{1}{4}$. Then again, these customers *were* right, in their way: right about the challenges of fractions and right about the subtlety of comparing numbers that possess infinite wardrobes of masks. Even those of us who recognize $\frac{1}{3}$ as larger than $\frac{1}{4}$ may struggle to sort, say, $\frac{3997}{4001}$ from $\frac{4996}{5001}$.

Of course, no pastry shop divvies their goods into 5001 pieces. But math does not bother with knives and crumbs. It serves its delicacies directly to the mind.

Decimals

I once drew a cartoon that was seen by 2 million people in a week, which is more than visit my blog in a typical year. Naturally, the joke wasn't mine. It was proposed to me by Howie Hua, the internet's friendliest math teacher, who tapped into a deep and primal conflict.

Fractions charm me. They're like a mom-and-pop bakeshop, happily slicing cakes however you wish. ("14 pieces out of 17? You got it, honey.") But I sometimes struggle to see the charisma of decimals. They're a coldhearted industrial operation, a factory of automated machinery that cuts cakes into 10 pieces. Always 10: no more, no less.

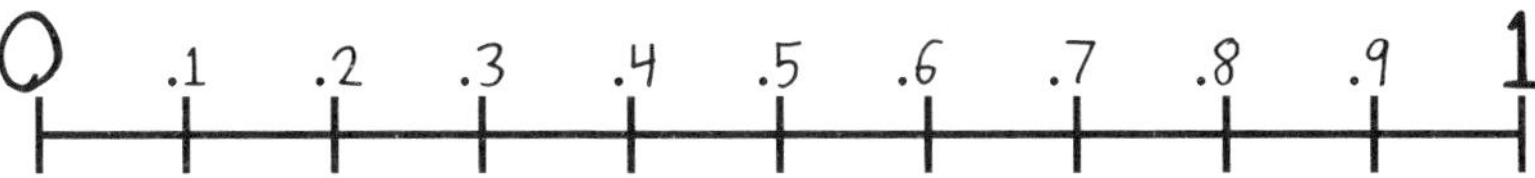

Need finer gradations? Slice each tenth into tenths.

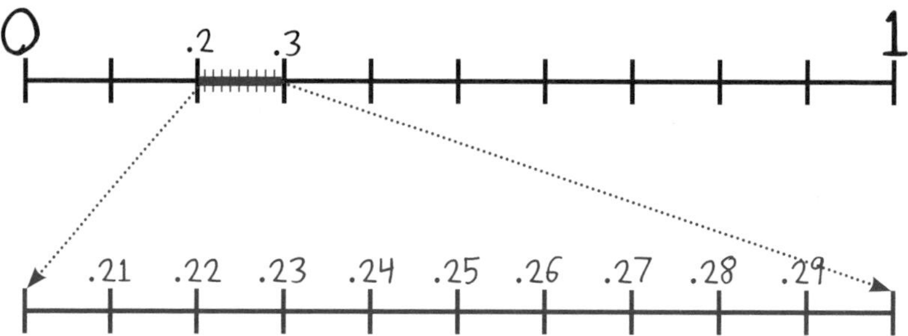

Even finer? Slice each of *those* into tenths.

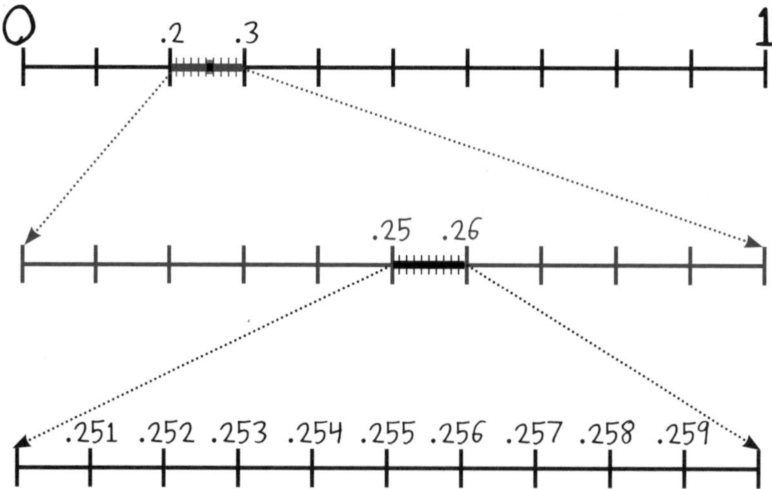

In this way, decimals extend the logic of the base ten system. Just as we build large numbers from tens of tens of tens, now we can specify small numbers using tenths of tenths of tenths. Indeed, mathematicians use "decimal" as a synonym for "base ten."

Tens all the way up, and tens all the way down.

I can't deny the advantages. Whereas comparing fractions can get hairy (is $^{17}/_{25}$ bigger or smaller than $^{54}/_{80}$?), comparing decimals is a hairless affair (clearly 0.680 is bigger than 0.675). Similarly, whereas fraction arithmetic requires nimble new algorithms (to add $\frac{3}{8} + \frac{2}{5}$, we must translate to $\frac{15}{40} + \frac{16}{40}$), decimal arithmetic does not (adding 0.375 + 0.400, we simply get 0.775).

But these strengths belie a terrible flaw: In the language of decimals, certain ideas are unsayable. There are thoughts with no words, numbers with no names.

Take ⅓: a simple, everyday quantity of which decimals cannot speak. One-third slips through the cracks of the machinery. It falls between 0.3 and 0.4, between 0.33 and 0.34, between 0.3333333 and 0.33333334. Though they come achingly close to ⅓, such approximations never quite reach it. To express ⅓ itself, we need a decimal of infinite length, an endless parade of 3's. If your hand cramps, or the notebook fills up, or the digits are truncated by the edges of the known universe—if the 3's ever, for any reason, cease—then you have not expressed the number ⅓, but a mere approximation thereof.

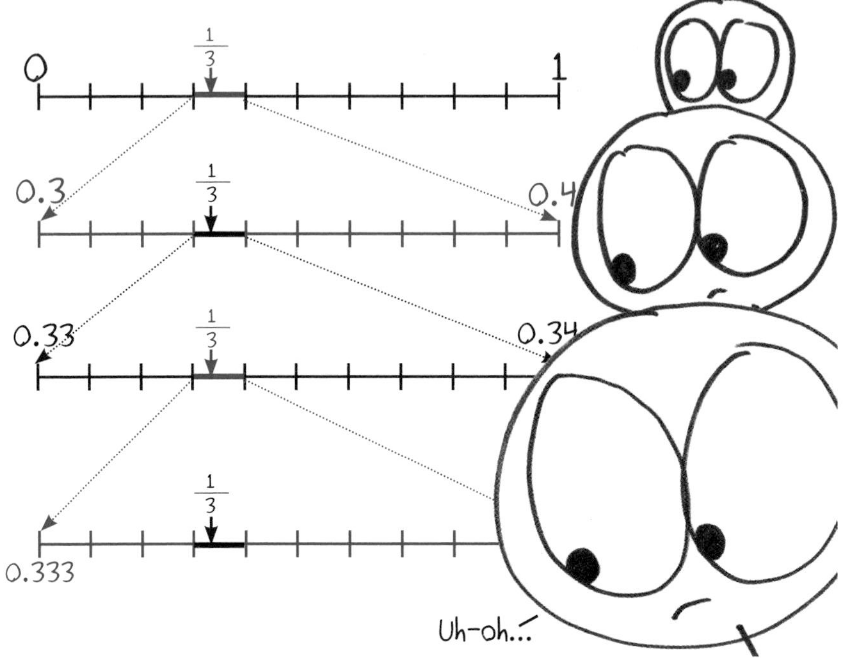

Nothing against approximations. Most days, good enough is good enough. For example, if you want ⅓ of a meter of floss, then 0.333 gets you to within the thickness of a credit card, 0.33333 to within the breadth of a hair, and 0.33333333333333333333333333333333333333 to within the smallest distance where "measurement" even makes sense. No reason to go further. Everyone accepts the limitations imposed by quantum mechanics—even flossers.

Still, it's funny that decimals have a reputation for precision. The opposite is true: Decimals are the language of approximation. Theirs is the mathematics of "Eh, close enough."

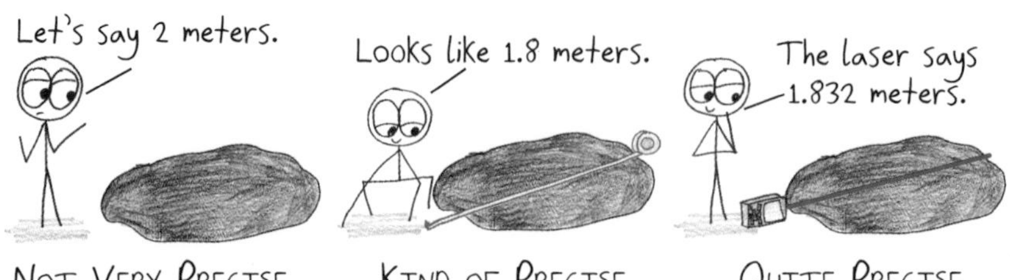

Anyway, since no decimal equals ⅓ and we'd rather not exhaust the planet's pens in a futile gesture at eternity, we take a more creative approach. That is, we cheat. We write $0.\overline{3}$ (pronounced "zero point three repeating") and call it a win.

HOW TO WRITE AN INFINITELY LONG REPEATING DECIMAL WITH ONLY FINITE PAPER AND PATIENCE

0. 3 3 3 3 3 3 3 3 3 3 3 3 and so on... → $0.\overline{3}$

0. 189 189 189 189 189 and so on... → $0.\overline{189}$

0. 4682 67 67 67 67 67 and so on... → $0.4682\overline{67}$

When I first taught 11-year-olds and assigned questions with solutions such as ⅔ or ⅚, I expected fractions as replies. I expected wrong. My students refused to give a fraction the final word and would instead write $0.\overline{6}$ (meaning 0.666666666...) or $0.8\overline{3}$ (meaning 0.83333333...). I tried to end this foolishness. "Imagine you speak two languages," I'd say. "One of them has a word for 'door.' The other does not: you just start saying 'Tuk-tuk-tuk-tuk-tuk...' and repeat that syllable forever. When you need to talk about doors, which language are you going to use?"

The classroom would erupt with a chorus of "Tuk-tuk-tuk-tuk," and I knew that Team Decimal had won this round.

Rounding

Strolling down the avenue, you meet a prophet. Flowing robes, glowing eyes, the whole deal. Allowed one query, you ask, "How long will civilization as we know it endure?"

The whispered reply: "1000 years."

You walk away somewhat consoled. A millennium is a long time, and anyway, there's no reason to expect the end in 1000 years on the dot. Maybe it'll be 1017 years, or 1194, or maybe the end will come so gradually that no single year is identifiable as—

Wait, I'm sorry. I've failed as a narrator. The prophet actually whispered something rather less consoling: "997 years, 119 days, 14 hours, and 33 minutes."

This grim thought experiment highlights a crucial element of mathematical grammar: rounding. To round is to communicate a number to a specified level of precision—say, the length in words of my book *Math with Bad Drawings* (82,771) to the nearest thousand. In this case, we find the thousands between which our number falls (82,000 and 83,000) and select the nearer of the two. (The halfway point 82,500 could go either way, but traditionally we round it up.)

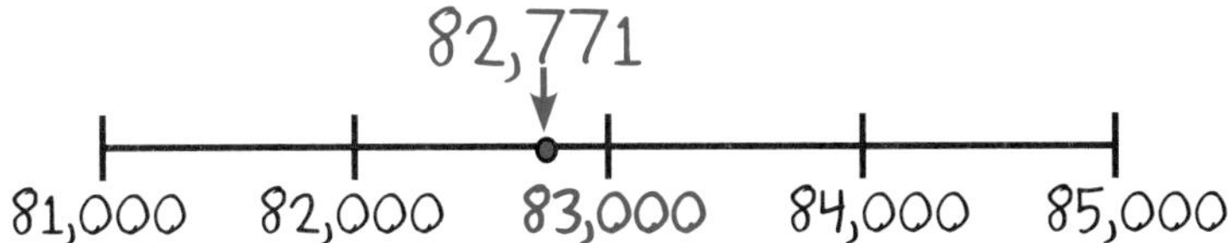

Rounding blurs details. An "83,000-word" book might in reality fall anywhere between 82,500 and 83,500 words. If we're rounding to the nearest 10,000, an "80,000-word" book could be as short as 75,000 words or as long as 85,000. That's quite a range, and it raises a question: What kind of snake-eyed embezzler would want to make the details blurry on purpose?

Well, in this mess we call reality, we blur details because the details are already blurred. As physicist Niels Bohr once said, "Never express yourself more clearly than you can think."

Sometimes a number is being put to a blurry purpose. For example, if you're wondering how many hours it'll take to read *Math with Bad Drawings*, then the word-by-word precision of 82,771 is unnecessary—or worse, deceptive. Reading speeds vary. It's like calculating a driving distance to the nearest tenth of a mile, not knowing whether you'll be moving at 25 miles per hour or 75.

Other times, the number emerges from a blurry measurement process—which is to say, any measurement process. No pipette is perfect. No stopwatch is sinless. When measuring a three-year-old's height, saying "93.7 centimeters" is a false claim to precision. More honest to say "94 centimeters" (or even just "90 centimeters," depending on how wriggly your toddler is).

In his book *Proofiness*, Charles Seife tells a joke about a guide at the natural history museum who informs visitors that the T. rex skeleton is 65,000,038 years old. "Wow," someone says. "How were they able to date it so precisely?" "Well," the guide explains, "it was 65 million years old when I started working here, and that was 38 years ago."

The blunder is obvious: treating a rounded number (65 million to the nearest million years) as if it were precise (65 million to the nearest year). It's a fantasy of perfect knowledge in a world that's imperfectly knowable. Seife calls this kind of error *disestimation*: the opposite of estimation.

Start looking, and you'll see disestimates everywhere. I grew up on the fiction that the human body temperature is 98.6°F. In truth, healthy temperatures range from 97°F (or lower) to 99°F (or higher). It's as if someone measured a temperature of 37°C, then converted into Fahrenheit—but somehow forgot that 37°C was rounded to begin with.

You can't specify healthy body temperature to the nearest 0.1°F. Healthy body temperature isn't that specific.

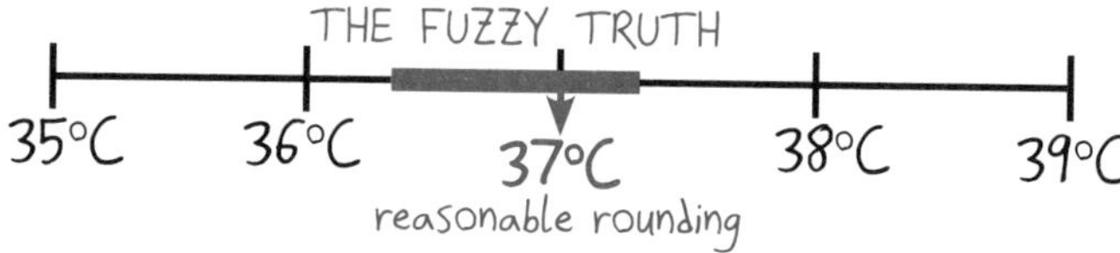

This brings us to the deepest cause of numerical blurriness: the inescapable blurriness of the world itself.

For example, what does it mean that *Math with Bad Drawings* is 82,771 words long? Does this figure include the endnotes? What about the words in the cartoons? Does an equation such as 6 + 4 = 10 count as one word or five? Different choices would produce different results: perhaps 82,775, or perhaps 82,894. Who's to say which number is right? Who's to say there's a "right" number at all?

If the value might fluctuate by hundreds, then the last three digits of 82,771 are meaningless. Rounding to 83,000 would better convey our true state of knowledge.

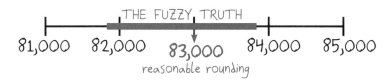

THE FUZZY TRUTH
81,000 82,000 83,000 84,000 85,000
reasonable rounding

Perhaps the silliest tic of modernity is that we expect precision on demand. I'm as guilty as anyone. I recall complaining once that the Olympics reports 100-meter dash times only to the nearest 0.01 seconds, rather than 0.001. "Can they not afford better clocks?" I cried. "Do they need me to buy them a $6 stopwatch?"

Then I realized: In 0.001 seconds, a sprinter travels barely 1 centimeter. Given ruffling shirts and waving hair, can we really specify a runner's location to within half an inch? The limiting factor isn't time but space—not our failure to measure but reality's failure to be measurable.

Rounding is blurry language, as befits a blurry world.

WORLD RECORD:
100(ish) meters in 9.58(ish) seconds

The need for rounding is why scientists of all stripes favor decimals over fractions. Each decimal effortlessly communicates its own level of precision: a 5.0-kilogram watermelon has been weighed to the nearest tenth of a kilogram; a 5.000-kilogram watermelon, to the nearest thousandth of a kilogram; and a 5.0000000000000000000000-kilogram watermelon, to the nearest carbon atom. Not so clear with fractions. Telling you an apple weighs 1/5 of a pound gives little indication of how reliable or wobbly my kitchen scale is. For a mathematician, fractions embody precision; they are a language of exactness, ill-suited to an inexact world.

Let us return, by way of closing, to our prophetic whisper: *997 years, 119 days, 14 hours, and 33 minutes.* Such precise language implies a precise event. Ours shall be no gradual decline, like the decay of the Roman Empire or the creeping irrelevance of rock music. The apocalypse can be dated down to the minute. Armageddon will come in a flash.

Or perhaps, if we're lucky, the prophet is only guilty of disestimation.

Large Magnitudes

With all of the glories that the farm had to offer—petting zoo, pumpkin patch, tractor ride—I could not fathom why four-year-old Frieda led us on a beeline to the corn pit. Really, kid? A pit full of corn? But I understood when I saw it: an area the size of several swimming pools, cordoned off with hay bales and filled waist-deep with more kernels than the eye could ken.

"That," I told my then two-year-old, summoning all of my worldly wisdom, "is a lot of corn."

How much? I had to know. Pacing off the perimeter and mumbling calculations, I estimated the total at 300 million kernels. Roughly the population of the United States. Frieda's mother, a poet, took a suitably poetic leap: "Hey, kids," she announced, "there's a kernel of corn here for everybody in the country. See if you can find yours."

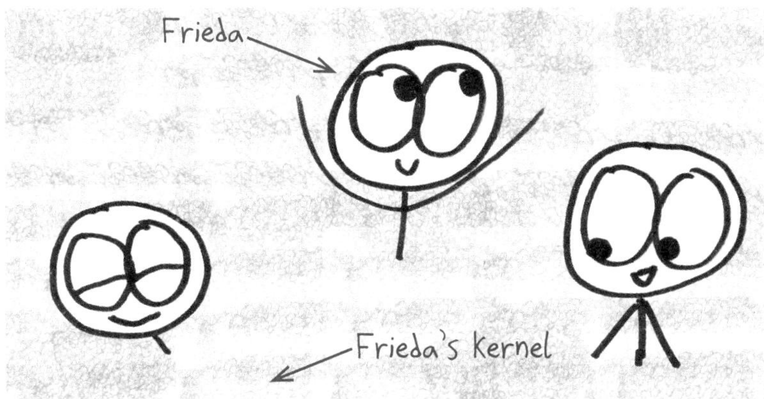

I love large numbers. I love imagining them. I love failing to imagine them. I love quantities so big it feels blasphemous even to name them.

Ancient cultures didn't always feel the same attraction. The larger Roman numerals ↂ (for 10,000) and ↈ (for 100,000) faded from regular use; for daily purposes, M (for 1000) was big enough. But we moderns live amid preposterous abundance: cities of millions, social networks of billions, exchanging bytes by the quintillions. We need stratospheric language for our stratospheric lives.

One crucial bit of vocabulary is the "exponent": a little superscript perched on another number's shoulder. It signifies "multiply together this many copies of the number." For example, 10^2 means 10 × 10 (or 100) while 10^6 means 10 × 10 × 10 × 10 × 10 × 10 (which comes out to 1 million). Thus, a small change to the exponent yields a big change to the result: 10^{19} is ten times larger than 10^{18}, and a million times larger than 10^{13}. These powers of 10 are sometimes called *orders of magnitude*.

NAME	POWER	DIGITS
One		1
Ten	10^{1}	10
One hundred	10^{2}	100
One thousand	10^{3}	1000
Ten thousand	10^{4}	10,000
One hundred thousand	10^{5}	100,000
1 million	10^{6}	1,000,000
10 million	10^{7}	10,000,000
100 million	10^{8}	100,000,000
1 billion	10^{9}	1,000,000,000
10 billion	10^{10}	10,000,000,000
100 billion	10^{11}	100,000,000,000
1 trillion	10^{12}	1,000,000,000,000
10 trillion	10^{13}	10,000,000,000,000
100 trillion	10^{14}	100,000,000,000,000
1 quadrillion	10^{15}	1,000,000,000,000,000

But, hey, why stop there? The Pacific Ocean holds 10^{20} gallons of water. Earth weighs 10^{25} pounds. The visible cosmos houses 10^{80} atoms. A game of chess can unfold in 10^{120} ways, and a game of go in 10^{500} ways. Dizzy yet?

Of course you are. No one can picture such things. Heck, we can't even handle 10^6 and 10^9. In a recent three-year span, the *Los Angeles Times* accidentally swapped the words "million" and "billion" on 23 occasions. Not to be outdone, the *New York Times* made 38 such errors. On Wall Street, similar screwups (such as the time a Bear Stearns trader accidentally tried to sell $4 billion in shares instead of $4 million) are known as "fat finger" mistakes. And don't forget the newly elected congressperson who disclosed a family asset worth $1 billion, then sheepishly amended the number to $1 million.

It's tempting to blame our language. "Million" and "billion" differ by just one letter; 10^6 and 10^9 by just one symbol; 1,000,000 and 1,000,000,000 by just three seemingly worthless zeroes. We could clarify the distinction by abandoning our fancy modern notations and reverting to plain old tally marks. A million tally marks would fill this book; a billion would fill about six bookcases' worth. Much harder to conflate. But—and perhaps this goes without saying—six bookcases of tally marks would be a pain to write. When it comes to big numbers, a little confusion is the price we pay for concision.

So how do we overcome this vertigo, this symptom that scholar Douglas Hofstadter dubs "number numbness"? It helps to get concrete. Pick a unit, any unit, and the million/billion difference is suddenly quite stark.

UNIT	MILLION	BILLION
Dollars	Buys a nice house	Buys a 100-story skyscraper
People	Population of San Jose, California	Population of Western Hemisphere
Seconds	About 12 days	About 32 years
Feet	Distance from NYC to Boston	Distance from NYC to the moon
Bytes	Audio file for "Eleanor Rigby" (first half only)	Audio files of all songs the Beatles ever released (two copies each)
Calories	Enough to feed a person for 16 months	Enough to feed a person for 1300 years

In his book *Innumeracy*, John Allen Paulos suggests memorizing a vivid image for each number. (Ten thousand: the approximate number of years it would take to transport Mount Fuji by truckload.) I say, go further. Memorize as many images as you can. Pick a single unit, a natural and meaningful unit, and then build a whole tower of images, one for each order of magnitude, climbing until the numbers stop making sense.

For example, take each number as a population:

PEOPLE	POPULATION OF . . .
1	A studio apartment
10	A two–family home
100	An apartment building
1000	A rural town
10,000	A suburb
100,000	A small city
1 million	A big city (e.g., Austin), or a small country (e.g., Djibouti)
10 million	A megacity (e.g., Bengaluru), or a medium country (e.g., Greece)
100 million	A large country (e.g., Egypt) or a large social network (e.g., Twitch)
1 billion	A huge country (e.g., India), or a huge social network (e.g., TikTok)
10 billion	The world, plus a bonus China
100 billion	Every human who has ever lived

Alas, long before we reach a trillion, these images begin to fray. I can visualize 10^5 as a huge stadium, and 10^6 as the crowd at a presidential inaugural, but my inner eye gives way at 10^7. Crowds of that size—the entire Paris metro area, or the population of the Dominican Republic, or the audience for a superviral video—are too diffuse to resolve into pictures. They're only facts. Inert. Abstract.

"There are 1,198,500,000 people alive now in China," Annie Dillard once wrote. "To get a feel for what that means, simply take yourself—in all your singularity, importance, complexity, and love—and multiply by 1,198,500,000. See? Nothing to it." As Dillard knows, the task is absurd. Yet we keep attempting to imagine billions, for the simple and compelling reason that we *are* billions.

Another unit worth exploring is the humble (or, depending on your politics, nefarious) dollar.

DOLLARS	VALUE OF . . .
$1	A candy bar
$10	A book
$100	A cheap doghouse
$1000	A high-quality doghouse
$10,000	A one-car garage
$100,000	A mobile home
$1 million	A mansion (or, in certain cities, a one-bedroom apartment)
$10 million	A public library for a midsized city
$100 million	A science building for a university

DOLLARS	VALUE OF . . .
$1 billion	A 100-story skyscraper
$10 billion	All the real estate in Gary, Indiana
$100 billion	All the real estate in Indianapolis
$1 trillion	All the real estate in Boston
$10 trillion	All the real estate in the UK
$100 trillion	All the real estate in the US
$1 quadrillion	All the things in the world

Again, imagination stumbles halfway up the ladder. I can grok $\$10^6$ (about two decades of earnings for the typical US worker) but not $\$10^7$ (two centuries for that same worker). I'm reminded of a line from Thomas Hardy's *Two on a Tower*: "There is a size at which dignity begins; further on there is a size at which grandeur begins; ..., further on, a size at which ghastliness begins." So if $\$10^5$ is dignity and $\$10^6$ is grandeur, then what is $\$10^7$?

Let us not finish with the ghastliness of obscene wealth. Join me instead on one final ascent, contemplating not the empty riches of capitalism but the rich emptiness of time.

YEARS AGO	BIRTH OF . . .
1	Current babies
10	Current fifth graders
100	My grandparents
1000	First people to see fireworks
10,000	First people to live in cities
100,000	First humans to speak language
1 million	First hominids to use stone tools
10 million	First ancestor of ours who is not also an ancestor of gorillas
100 million	First mammals
1 billion	First multicellular organisms
10 billion	First galaxies

What to make of these magnitudes? Heck if I know. Just as there are poems I relish reciting even if I can't grasp their meaning, so, too, do I love running my hands through the kernels of time and muttering to no one in particular, "Now *that* is a lot of corn."

Scientific Notation

When I lived in the United Kingdom, I kept a tally of minor differences in mathematical language. Little disputes between the dialects of my old home and my new one. Here are a few favorites.

	US CALLS IT . . .	UK CALLS IT . . .	WHO'S RIGHT?
	Trapezoid	Trapezium	UK. "Trapezoid" stems from an error made by some dude in the 1700s.
3^7 or 10^{-5}	Exponents	Indices	US. Especially when Brits give the singular not as the correct "index," but the lexical nightmare "indice."
Mathematics	Math	Maths	They're interchangeable, and anyone who argues otherwise is a pedant.

But one difference perplexed me above all others:

	US CALLS IT . . .	UK CALLS IT . . .	WHO'S RIGHT?
$6.02 * 10^{23}$	Scientific notation	Standard form	TBD in this chapter

Call me jingoistic, but I'm inclined to award this point to the US. To me, "standard form" is "8,200,000,000," or perhaps "8.2 billion." Show $8.2 * 10^9$ to your aunts and uncles, and they may say "Ah, the notation of scientists," but probably not "Ah, the standard form in which to write a number."

Still, I get the idea. Scientific notation, while not standard in news outlets or family gatherings, perhaps should be.

Huge numbers can be hard to tell apart. The gaps between 80000000 (the population of Turkey), 800000000 (the population of Europe), and 8000000000 (the population of Earth) are chasmic yet practically invisible to the naked eye. You've got to count the digits, zero by painstaking zero. To ease the headache, we use commas to break the symbols into triplets: 80,000,000 for Turkey, 800,000,000 for Europe, and 8,000,000,000 for the world.

But past a certain length, the commas are little relief. Try your eye on 800,000,000,000,000,000,000,000,000,000,000,000,000,000. Is that 800 duodecillion, or 800 tredecillion? And while I'm asking, what the heck is a tredecillion? Clearly, we need a fresh approach.

Enter scientific notation. The concept is simple: when naming a number, first name the order of magnitude, and then how many of that magnitude we have.

NUMBER	MAGNITUDE	HOW MANY?	AND THUS . . .
5,880,000,000,000 (miles in a light-year)	Trillion (10^{12})	5.88 (trillions)	$5.88 * 10^{12}$
340,000,000 (US population)	Hundred million (10^8)	3.4 (hundred millions)	$3.4 * 10^8$
602,200,000,000, 000,000,000,000 (atoms in a mole)	Hundred sextillion (10^{23})	6.02 (hundred sextillions)	$6.02 * 10^{23}$

It helps to read scientific notation from right to left. What really matters is the magnitude. Then, for extra detail, you can then check the multiplier. Thus, $3.4 * 10^8$ is in the hundreds of millions; and if you want to get specific, it's 340 million.

This approach isn't just for scientists. My stepmother, Lark, a lawyer by training, a nonprofit CEO by vocation, and "not a math person" by self-avowal, once explained how she understands budget items. "First, we need to know what kind of number we're talking about," she told me. "Tens of thousands of dollars? Or just a few thousand? Or a few hundred thousand? Then we can say more precisely how many—say, 3000, or 8000."

When I told her that's exactly how mathematicians think, she glared as if I'd said not "mathematicians" but "hired assassins." Apologies to Lark, but the connection stands. Scientific notation is just the formalization of common sense: a standard form for numerical thoughts. Maybe the Brits win the point after all.

But wait—I've told only half the story.

Yes, scientists traffic in telescopes and trillions. But they also deal in microscopes and millionths. How should we speak of small things and tiny numbers?

Every tiny number is the mirror image of a huge one: thousand and thousandth, million and millionth, billion and billionth. This goes all the way to the top (and, of course, to the bottom). When 17th-century mathematician John Wallis coined ∞ as a symbol for the infinitely large, he also proposed a notation for the infinitely small: $\frac{1}{\infty}$. (Pro tip, kids: use this notation to infuriate your algebra teacher!)

We bring this same symmetry to the powers of 10. If the positive powers are big, then the negative powers must be small. If 10^{12} is a trillion, then 10^{-12} is a trillionth.

Power	Product	Digits	Name
10^3	10×10×10	1000	thousand
10^2	10×10	100	hundred
10^1	10	10	ten
10^0	1	1	one
10^{-1}	1/10	0.1	tenth
10^{-2}	1/10×10	0.01	hundredth
10^{-3}	1/10×10×10	0.001	thousandth

Does this make sense? Strictly speaking, no. We defined 10^5 to mean $10 * 10 * 10 * 10 * 10$. You can't multiply together –5 copies of 10, so in that respect, 10^{-5} is nonsense.

Still, it is pleasing and symmetrical nonsense. If 10^2 means repeated multiplication, then shouldn't 10^{-2} mean repeated division? Stepping up a magnitude (from 10^2 to 10^3) means multiplying by 10. Stepping down a magnitude (from 10^3 to 10^2) means dividing by 10. Just keep stepping down (10^1 to 10^0 to 10^{-1}) and keep dividing by 10 (10 to 1 to 0.1), and the negative powers sort themselves out.

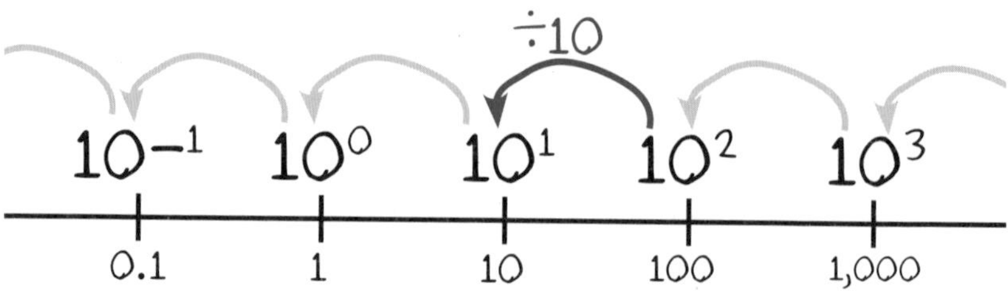

These tiny scales are vital for scientists, because, in a sense, there is more reality "below" us than "above." The meaningful magnitudes of space run from 10^{26} meters (the width of the visible universe) down to 10^{-35} meters (the Planck length). Similarly, the meaningful magnitudes of time run from 10^{-44} seconds (the Planck time) up to 10^{18} seconds (the span since the big bang). Either way, from our vantage point, the universe has more small magnitudes than big ones.

POWER	NAME	DIGITS	SECONDS FOR . . .	METERS FOR . . .
10^{0}	1	1	Beat of a human heart 	A toddler 
10^{-1}	100 *milli-*	0.1	Blink of an eye	A hummingbird
10^{-2}	10 *milli-*	0.01	Flap of a fly's wings	A coffee bean
10^{-3}	*milli-*	0.001	Pop of a bubble 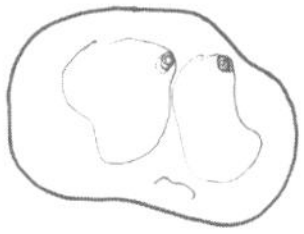	A large sand grain
10^{-4}	100 *micro-*	0.000 1	Sound to travel 1 inch	A human egg
10^{-5}	10 *micro-*	0.000 01	Bullet to travel half an inch	A white blood cell
10^{-6}	*micro-*	0.000 001	Flash of the fastest strobe lights available 	A particle of clay

POWER	NAME	DIGITS	SECONDS FOR . . .	METERS FOR . . .
10^{-7}	100 *nano-*	0.000 000 1	Time between two high-frequency radio waves	A virus
10^{-8}	10 *nano-*	0.000 000 01	Step of a nuclear reaction	An ice crystal
10^{-9}	*nano-*	0.000 000 001	Light to travel 1 foot	A carbon nanotube
10^{-10}	100 *pico-*	0.000 000 000 1	Light to travel 1 inch	An oxygen atom

While negative exponents have the minor disadvantage of not making sense (at least, not the same kind of sense that positive exponents make), the advantage is bigger: They let us express tiny numbers with the same clarity and concision as huge ones. In particular, they let us extend scientific notation down into the quantum realm.

NUMBER	MAGNITUDE	HOW MANY?	AND THUS . . .
0.000 03 (meters across a skin cell)	hundredths of thousandths (10^{-5})	3 (hundredths of thousandths)	$3 * 10^{-5}$
0.000 000 24 (meters across a virus)	tenths of millionths (10^{-7})	2.4 (tenths of millionths)	$2.4 * 10^{-7}$
0.000 000 000 17 (meters across a gold atom)	tenths of billionths 10^{-10}	1.7 (tenths of billionths)	$1.7 * 10^{-10}$

As I said, the cosmos only goes so small. At 10^{-35} meters, space hits bedrock. At 10^{-44} seconds, time bottoms out. Beyond a certain scale, the universe can be sliced no finer. That's the meaning of "quantum": A discrete unit. The smallest possible amount.

But with numbers, the slicing and dicing need never stop. Just as a thousand can be split into hundreds, so can 10^{-44} be split into 10^{-45} and then into 10^{-46}, and so on, down to $10^{-96,782}$ and beyond. There is no quantum level, no rock bottom, no final frontier. Numbers form a continuum, infinitely divisible.

In that sense, neither the Yanks nor the Brits have this one quite right. Expressions such as 10^{-500} are neither standard nor scientific. They're flights of pure mathematical whimsy.

Irrational Numbers

Every March 14, the mathematical world observes its favorite holiday by canceling lessons, gorging itself on pie, and reciting the decimal expansion of math's most revered constant. Young and old, pure and applied, algebraist and analyst, we all gather to celebrate Pi Day.

Except... Okay, I take that back. *Most* of us celebrate. A few grinches prefer to grumble about it.

We'll come back to them, but first: What is pi? It's the number of *diameters* (the distance across a circle) you'd need to equal the *circumference* (the distance around a circle). Roughly speaking, three *acrosses* equal one *around*.

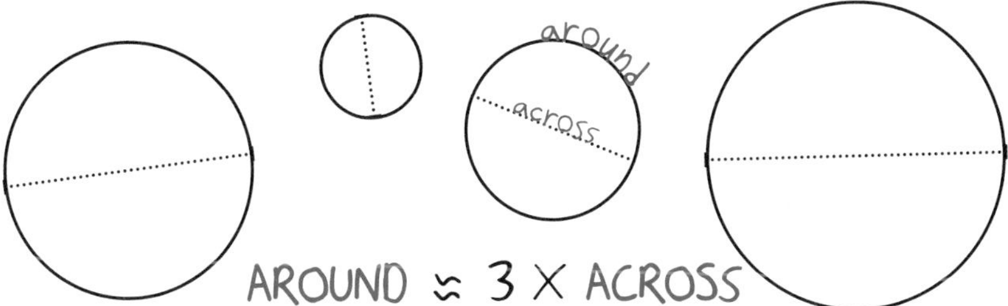

More specifically, the around is about 3.14 times the across.

Even more specifically, it's about 3.141 592 653 589 793 times.

Hyperspecifically, it's 3.141 592 653 589 793 238 462 643 383 279 502 884 197 169 399 375 105 820 974 944 592 307 816 406 286 208 998 628 034 825 342 117…

I can get more specific, but never specific enough, because this number, known as π or pi, is irrational. As in, not a ratio. No fraction can express it. No decimal either. We can't even pull a trick like when we wrote $0.\overline{3}$ for $\frac{1}{3}$, because π's digits don't settle into a repeating pattern.

Around every corner is a new, never-before-seen digit sequence. I have heard 12-year-olds recite a hundred digits from memory. And that's nothing compared to the 70,030 digits reeled off by Suresh Kumar Sharma in October 2015, over the course of 17 (presumably agonizing) hours.

So why does this drive mathematicians batty? With apologies to Dr. Seuss, I find it easiest to express their lack of holiday cheer through rhyme.

Every kid in the classroom liked Pi Day a lot.
But the Grinch in the faculty lounge
did not!
PI DAY
PARTY!
3/14

This, to the Grinch, was the dumbest of seasons,
and asked or unasked, he would rattle off reasons:
"In the US, it's said the date 3 slash 14
is March the 14th. That's what 'Pi Day' must mean.
But the rest of the world gives the day, then the month!
So 3 slash 14 is the 3rd of Kerplumph,
a month, I assure you, that does not exist.
And that is just one of the gripes on my list.

This horrible day that transfixes the nation,
is built on a lousy approximi-zation.
22 over 7 comes closer to pi.
So let's wait till that date, out in mid-late-July.

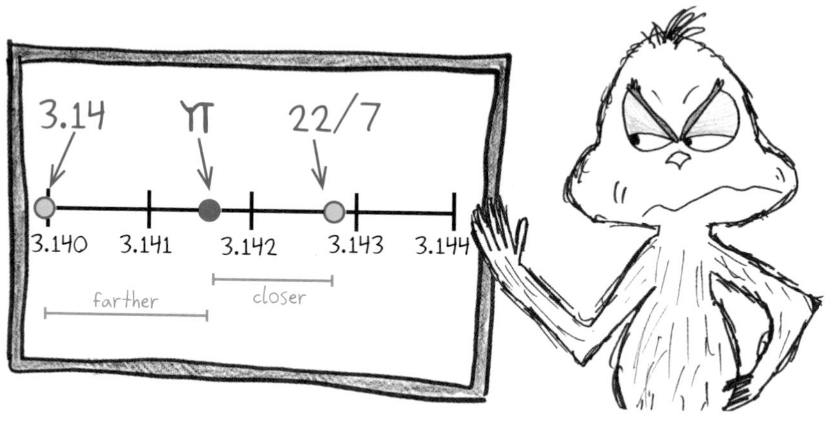

Besides which, old pi is an artifact now.
The cool mathematicians all venerate tau.

And have I brought up
how it eats me alive,
the 3 point 1 4 1 5 9 2 6 5?
How it just rambles on,
and it fills me with hate.
3 5 8 9 7 9 3 2 3 8!
All those meaningless digits!
The hot wasted breath!
Oh, it bores me to tears.
No, it bores me to death.

I've said it before, and I'll say it again:
Those numbers mean nothing outside of base ten.
Besides which, the digits are ones no one needs.
Past 30 or 40, they're useless as Thneeds.
For a circle a million light-years in width,
your digits could stop at the mere 25th,
and still you could calculate out the circumference
down to the span of a nano-non-umference,
a distance one-thousandth the width of a hair.
For digits past that, I quite simply don't care.

3.141592653589793
2384626433832795
0288419716939937 5
10582097494459 23
0781640628620899
86280348253421...

I'm speaking the truth
when I firmly assert:
Pi Day's an excuse
to swap math for dessert."

Setting aside calendrical arguments to focus on mathematical ones, I see two strong points in the grinches' favor.

First, irrational numbers are not rare. Throw a dart at the number line, and you'll hit one. If what we care about is irrationality, then we might as well replace Pi Day with $\sqrt{17}$ Day on April 12, or $\frac{3e}{7}$ day on January 16. Yes, π is more important than these numbers, but dwelling on its irrationality is like centering MLK Day celebrations on Martin Luther King Jr. having been 5'7". Kind of missing the point.

Second, even if irrational numbers *were* rare, then memorizing their digits would still be a silly pastime. You can typically round π to 3.14159, or 3.14, or even 3. For all practical purposes—and even for impractical ones—π might as well be rational.

That said, few grinches follow this logic to its terrible conclusion: *irrationals don't exist.*

Infinite precision is impossible. No ruler, scale, or stopwatch can give you unlimited decimal places. Sooner or later, you must round.

And once you round, the irrational number is gone, replaced with a boringly rational approximation.

So in what sense do irrationals exist, except in our imaginations?

For 364 days of the year, we must accept the dreary reality that beyond the first handful, an irrational's digits are empirically (if not existentially) meaningless. But one day each year, the world indulges our fantasy that irrational numbers exist. For one day, the world stops to admire a number that escapes description, a noun that can never be uttered.

Plus, we get to devour fistfuls of pecan pastry. What's not to love?

And what happened then?
In some circles, they say
the Grinch's small heart
grew three sizes that day.
And elsewhere, they say
that it grew a bit more:
maybe 3 point 1 sizes,
or 3 point 1 4...

Infinity

"Only three things are infinite," the writer Gustave Flaubert once noted, "the sky in its stars, the sea in its drops of water, and the heart in its tears."

Wrong, wrong, and wrong.

There are at most 10^4 stars visible in the night sky, with roughly 10^{11} in the galaxy, and perhaps 10^{24} in the knowable cosmos. In any case, finite. As for water, Earth has about 10^{25} drops. Again, finite. And on the subject of tears, I have two lessons for Gustave. First, teardrops come from tear ducts, not from cardiac muscle. Second, however numerous they may be, they are decidedly finite in number.

As a matter of fact, no things are infinite. Likewise, infinity is no thing. Infinity is more of a gesture, akin to saying "Go west and never stop." When mathematicians say that something "goes to infinity" or "becomes infinite," they really mean that it grows and grows and grows, past millions and billions and trillions, beyond any ceiling you can imagine. But it is always, at every step, finite. It never "becomes" infinite, because infinite isn't a thing you can become.

Infinity is not a destination but a direction.

So there you have it. I've gestured in the direction of infinity, which is all that Gustave was really trying to do, and, indeed, all that one *can* do. You may now safely forget the whole idea.

Infinity

Still here, are you?

Fine. "What *is* infinity?" you ask. Well, it's right there in the word: *not finite*. The synonyms are similar: "boundless," "limitless," "endless," "fathomless." We define "infinity" by what it is not (finite), what it lacks (bounds, limits), what it never does (end), and what it can never be (fathomed). We cannot speak of infinity on its own terms, but only as a kind of negation: the opposite of what exists, the repudiation of all that we know.

Please, I beg you, leave it at that. Dabble no more in these riddles and curses. Move on with this precious, finite day.

Infinity

Oh, all right, you want to summit the infinite? It's a fool's errand, but as a card-carrying fool, I feel compelled to join you. "Perhaps universal history," wrote Jorge Luis Borges, "is the history of a few metaphors." Infinity is one of those, a metaphor for unspeakable vastness. 16th-century scholar Giordano Bruno, in striving to elucidate the structure of the Copernican cosmos, hit upon the image of an infinite sphere, "the center...everywhere and circumference nowhere," this being the same image by which centuries of theologians had articulated the nature of God. Yet infinity, for all its vastness, sits close at hand. The obverse of the infinite is the infinitesimal, and therein lies the trouble, for the infinitesimal has plagued mathematics ever since the 5th century BCE, when the philosopher Zeno first laid out his wicked paradoxes. That's why the generations after John Wallis resisted his uncanny fraction $\frac{1}{\infty}$, sensing that it was only a name for a confusion. The infinitesimal—and thus, the infinite—came to signify the frontier where logic dissolves, the paradoxical site of an everything that comes from something and a something that comes from nothing. Infinity is the Jungian serpent that devours its own tail, a thought that cannot truly be thought—or so it was until the late 19th century, when Georg Cantor clapped infinity in the irons of rigorous mathematics. He caged the infinite within the language of sets and collections, bending the illogical to obey his logic (or perhaps vice versa). Through the bars of the cage, we could see infinity clearly now; we could see that a doubled infinity grows no larger, that a halved infinity grows no smaller, and that if infinite drawers each hold a marble, and each marble is replaced with infinite marbles, then all of the infinite infinitudes of marbles can still fit in the original drawers, one marble in each. Cantor revealed that all the paradoxes of infinity were not paradoxes at all, but elementary facts about an inconceivable thing, the features of a concept that our logic and language have grasped but our imagination never shall. Cantor taught us that there are smaller and larger infinities, a whole hierarchy of infinities, itself infinite in scope, and he dubbed these infinities $\aleph_0$ and $\aleph_1$ and $\aleph_2$ after the Hebrew letter *aleph*, which Jorge Luis Borges (there he is again) would later borrow for the title of a story that evokes infinity better than any mathematics I know. It is a brief story about a point in space: a point that contains all other points, so that the entirety of creation is visible through a crack in the floorboards. "The Aleph" culminates in a single monolithic paragraph describing the narrator's glimpse of the titular object: "I saw the teeming sea; I saw daybreak and nightfall; I saw the multitudes of America; I saw a silvery cobweb in the center of a black pyramid…I saw the circulation of my own dark blood; I saw the coupling of love and the modification of death; I saw the Aleph from every point and angle, and in the Aleph I saw the earth and in the earth the Aleph and in the Aleph the earth; I saw my own face and my own bowels; I saw your face; and I felt dizzy and wept, for my eyes had seen that secret and conjectured object whose name is common to all men but which no man has looked upon…"

÷10
10⁻¹ 10⁰ 10¹ 10² 10³
0.1 1 10 100 1,000
π ≈ 3.14
around
across
"857"
-(-(-(-1)))
≈ 6.28
y=mx+b
simple.
1/4 + 1/6 =
$2.50!
THE FUZZY TRUTH
96°F 97°F 98.6°F 100°F
125
124
249
∀ε>0,
∃δ>0:
x−a|<δ→
f(x)−L|<ε
ax²
+++
CHOCODOLLAR
unreasonable rounding
3.14 π 22/7
3.140 3.141 3.142 3.143 3.144
10×10
1/10×10×10
17 × whatever
1.414ish
PI DAY PARTY!
3/14
3.4104398
2.5899496
6.0003894
5×3
multiplication
4
7
$10
farther
closer
2573
× 389
3 + −1.78
division by 1/2
17 34
division by 2
0 5 10 15 20 25 30 35 40 45 50
repeated addition
5⁵⁵
5↑↑3
repeated exponentiation
tetration
λ
6
2
4^{7/2} = ???
NOOOOOO
a ÷ b/c = ac/b
conversation blender nightclub thunder
40 50 60 70 80 90 100 110 120
×1000
x(x+x²)?
x²+x³!
8÷2(2+2)
8÷2(4)
8÷8
1
8÷2(4)
4(4)
16
x²=3x+10
K=½mv²
If A<B and B<C
then A<C
x = −b ± √(b² − 4ac) / 2a
DONUTS
44.9° North
93.2° West
7 PIZZAS
10
6
ax² = bx + c
ax² + bx = c
ax² + c = bx
ax² = bx
ax² = c
bx = c
1/3 sleep deprivation,
2/3 game theory.
100%
95% to 99.9%
80% to 95%
60% to 80%
40% to 60%
20% to 40%
5% to 20%
0.1% to 5%
0%
AROUND ≈ 3 × ACROSS
OOOOH!
17
34
noisy
11
4n+2
"1000"
1 −1 −(−1) −(−(−1))
5↑↑3 Wow!
5+3 5×3 5³
30×20 30×7 8×20 8×7
600 + 210 + 160 + 56 = 1026
No! Silence your nonsense!
82,771
What is lim_{x→2} (x²−4)/(x−2)?
What's 46 + 28?
27
.1 .2 .3 .4 .5 .6 .7 .8 .9 1
see... calculating... calculating...

÷10
10⁻¹ 10⁰ 10¹ 10² 10³
0.1 1 10 100 1,000
π ≈ 3.14
"857"
-(-(-(-1)))
τ ≈ 6.28
y=mx+b
around
across
1/4 + 1/6 =
Ah, simple.
125
+ 124
249
∀ε>0,
∃δ>0:
x-a|<δ→
f(x)-L|<ε
ax²
+++
THE FUZZY TRUTH
96°F
97°F
98.6°F
unreasonable rounding
CHOCODOLLAR
10X10
1/10X10X10
3.14 π 22/7
3.140 3.141 3.142 3.143
17 × whatever
1.414ish
PI DAY PARTY!
3/14
3.4104398
+ 2.5899496
6.0003894
5×3
multiplication
4/7
farther
closer
10
2573
× 389
division by 1/2
17
34
division by 2
3 + -1.78
???
0 5 10 15 20 25 30 35 40 45 50
repeated addition
5⁵⁵ → 5↑↑3
repeated exponentiation
tetration
λ
NOOOOOO
4⁷/² = ???
6
2
whisper conversation blender nightclub thunder
20 30 40 50 60 70 80 90 100 110 120
×1000
x(x+x²)? x²+x³!
8÷2(2+2)
8÷2(4)
8÷8
1
x²=3x+10
K=½mv²
If A⊂B and B⊂C
then A⊂C
x = -b ± √b² - 4
2a
DONUTS
7 PIZZAS
10/6
44.9° North
93.2° West
1/3 sleep deprivation,
2/3 game theory.
ax² = bx + c
ax² + bx = c
ax² + c = bx
ax² = bx
ax² = c
bx = c
Guaranteed 100%
Almost certain 95% to 99.9%
Very likely 80% to 95%
Probable 60% to 80%
Maybe 40% to 60%
Possible 20% to 40%
Unlikely 5% to 20%
Almost impossible 0.1% to 5%
Impossible 0%
AROUND ≈ 3 X ACROSS
OOOOH!
17/34
noisy
11
"1000"
3
5
1 -1 -(-1) -(-(-1))
5↑↑3 Wow!
30×20 30×7 8×20 8×7
600 + 210 + 160 + 56 = 1026
5+3 5×3 5³
No! Silence your nonsense!
82,771
What is lim x→2 (x²-4)/(x-2)?
0 .1 .2 .3 .4 .5 .6 .7 .8 .9 1
What's 46 + 28?
27
let's see... calculating... calculating...

VERBS

The Actions of Arithmetic

A verb is a word of action. Nouns are things; verbs are what they do. Rabbits (*noun*) run (*verb*). Prices (*noun*) jump (*verb*). Demons (*noun*) slumber (*verb*). Writers (*noun*) ramble (*verb*), if you (*noun*) see (*verb*) what I (*noun*) mean (*verb*).

When it comes to math, what constitutes a verb? In a word, "operations." The four most familiar are addition (+), subtraction (–), multiplication (∗ or × or, in some settings, the absence of a symbol), and division (÷ or /). Once upon a time, people carried out these operations with a physical abacus, sliding around pebbles (*calculi*) to find answers (*calculate*). Eventually, their successors took up paper and ink, sliding around not rocks, but written marks. Today, we increasingly rely on computers, sliding around neither rocks nor marks, but electrons.

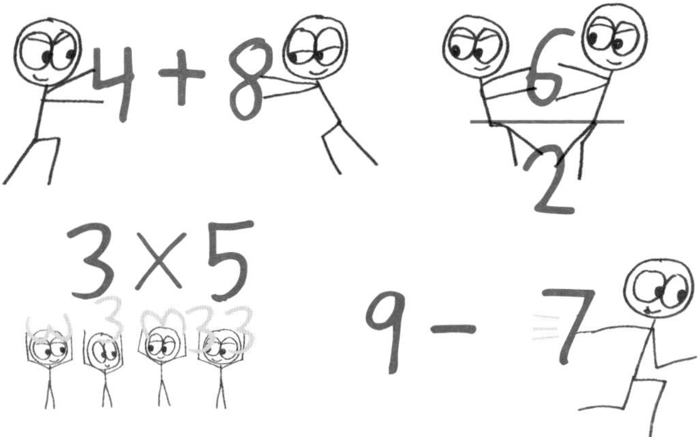

The story of calculation does not end there. Numbers are like potatoes: they lend themselves to many preparations. Witness the operations of squaring, cubing, roots, exponents, logarithms, and more. Before "computer" was a technology, it was a job description, referring to people (mostly women) who carried out these operations en masse, the numbers moving between them like products down an assembly line.

All of these are the verbs of the mathematician: the things we do to numbers.

In this section, we'll study some of these operations. We'll also ask how they relate to one another. How do these verbs fit together into a language?

First, the actions tend to come in opposite pairs: addition and subtraction, multiplication and division, squaring and taking square roots, exponentiation and logarithms. Such pairs are *inverses*, processes that undo one another. Perform one inverse after the other (lock the door, then unlock it), and you wind up where you started (with an unlocked door).

Second, the operations belong to a kind of hierarchy. Count repeatedly, and you get *addition*; add repeatedly, and you get *multiplication*; multiply repeatedly, and you get *exponentiation*. We'll see that mathematical operations are not read from left to right; they're read from most powerful to least powerful. The hierarchy structures the language.

So far, so good. But eventually, we'll have to confront an uncomfortable truth. These actions aren't really verbs at all.

Consider 2 + 3. If + is a verb, then who is the subject carrying out the addition? Neither 2 nor 3 performs any action; those nouns just sit there being nouns. *You're* the one who adds, but you're not a part of mathematical speech. Thus, in a strict grammatical sense, 2 + 3 isn't actually a sentence, because + isn't actually a verb. It's just a noun phrase—"two plus three"—something like "a cat and a bird" or "a dog with a bone." The + symbol is more like a conjunction (2 *and* 3) or a preposition (2 *with* 3).

This seemingly minor technical point turns out to be quite the opposite: a technical point of language-changing, mind-altering importance.

That comes later. For the time being, we'll treat + and – as math's elementary verbs. But first, there is a verb even more elementary, one as unconscious and fundamental as your heartbeat, and one that (like a heartbeat) requires a trained expert to inspect properly…

Increments

One morning I watched my sister, Jenna—a math teacher herself, and a better one than I am—greet kindergartners. "Good morning!" she'd say. "Tell me, how old are you?"

"Five!" they'd reply.

"So," Jenna would say with a gleam in her eye, "how old will you be next year?"

This was not a question that I deemed promising. Surely, five-year-olds know what comes after 5. Indeed, as I expected, many yelled out a prompt "Six." But here is why Jenna is the pro, and I am the little brother: Several children surprised me by reliving their whole chronology to date. To answer the question, they needed to recite all the prior numbers, just as you or I might hum the ABC's to recall whether Q comes before R.

"One, two, three, four, five…six. I'll be six!"

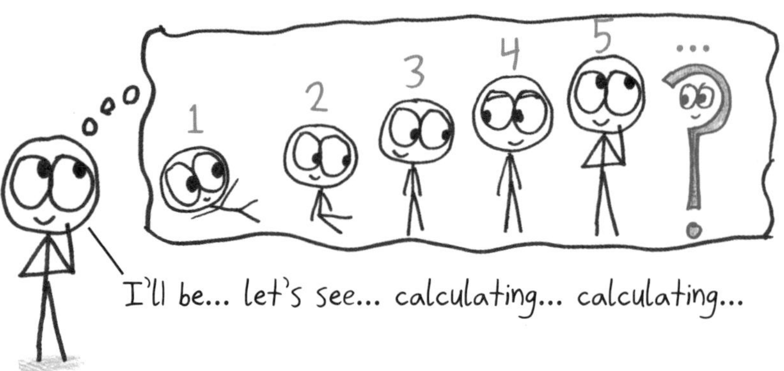

Operations are the things we do to numbers: dividing, exponentiating, taking cube roots, or even something as humble as adding. But, if you'll indulge a philosophical moment, are we really doing anything at all? It's a bit peculiar to say that I add 4 and 3 to create 7; 4 and 3 simply *are* 7, irrespective of my efforts. When I multiply or divide or take a logarithm, the result equals whatever it equals, regardless of my labors. I do not, in a strict sense, *change* the numbers, but merely *discover* or *reveal* them. Operations act not upon the quantities themselves, but only upon our understandings of them.

Then again, in a plainer and less philosophical sense, we act upon numbers all the time. As any schoolchild will tell you, that's what math is all about.

This brings me back to Jenna's inconspicuous question, which quietly engaged the kindergartners in an elemental act of mathematics: *incrementing*. That is, going from one number to the next. Incrementing is pretty simple, but only in the same deceptive sense that atoms are simple. It is the building block from which all other operations are built.

For example, what is addition if not a compact way to increment several times in a row?

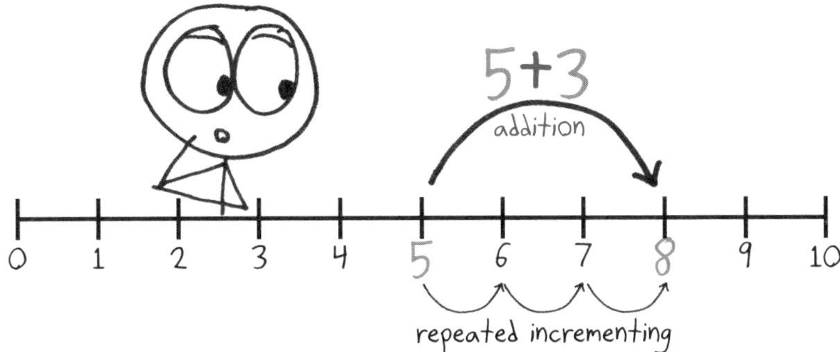

And what, then, is multiplication if not repeated addition, which is to say, repeated repeated incrementing?

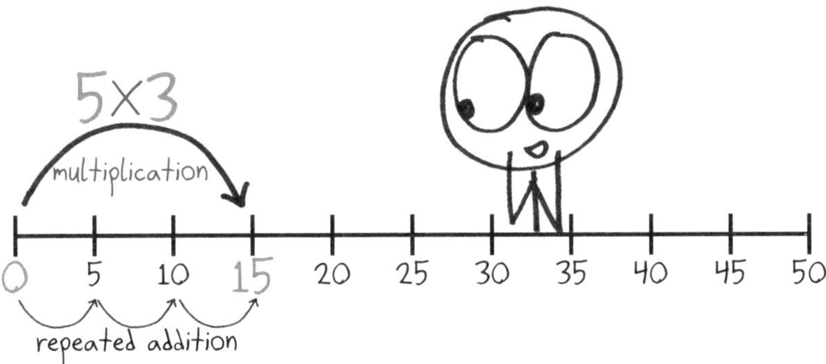

As for exponentiation, well, isn't that just repeated multiplication, i.e., repeated repeated addition, i.e., repeated repeated repeated incrementing?

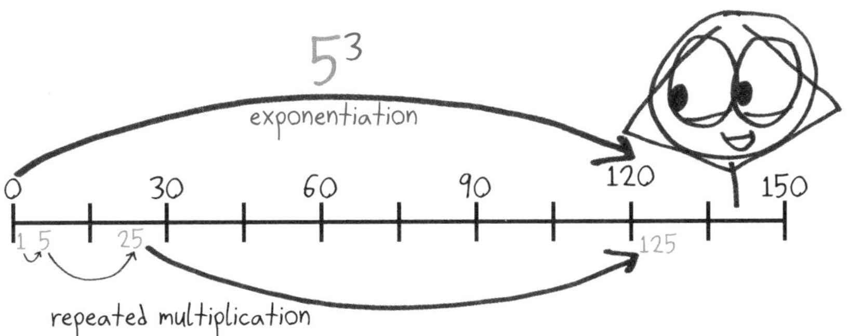

Why stop there? Repeated exponentiation creates a fourth-tier operation called *tetration* (from *tettares*, meaning "four"). We politely avoid it in school math because it yields numbers of uncouth size. Whereas the addition $5 + 3$ gives 8, and the multiplication 5×3 gives 15, and the exponentiation 5^3 gives 125, the tetration $5 \uparrow \uparrow 3$ yields an uncivilized number more than 2000 digits long.

Yet this mind-boggling operation with stupefying results is, at bottom, nothing more than repeated repeated repeated repeated incrementing.

I'll admit this is a reductive scheme. It is a bird's-eye view of operations, a way of capturing them all on a single map. In the process, it loses the particularity that makes each operation special. Describing tetration as "repeated incrementing" is like describing one's husband as "a collection of particles." Not wrong, but not necessarily a sign of a healthy marriage.

I prefer to treat operations like cities, each with its own local flavor and culture. The hidden geometry of squaring; the dual nature of division; the linguistic puzzles of square roots. An operation is the unique expression of a special truth—even the operation that prefigures all the others, the operation so elementary we barely recognize it as such, the operation that Jenna's keen question spotlighted.

For some of Jenna's kindergartners, incrementing pushed the limits of their powers. But other students—those who replied with an immediate "Six"—had already begun to act upon numbers, to wrestle them into more meaningful forms.

To *operate* on them.

When it comes to intellectual growth, moving from "One, two, three, four, five, six" to simply "Five, six" might seem like a marginal step. But, hey, what is a journey if not a succession of increments, a step repeated and repeated and repeated and repeated and repeated...?

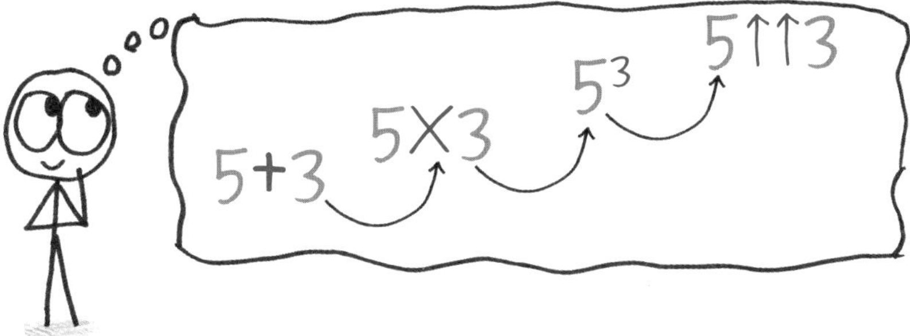

Addition

In my childhood, back when money was made of metal and paper instead of software and lies, I learned that two quarters add up to 50 cents: $25 + 25 = 50$. I loved knowing this. Those towering numbers made other facts feel puny; I now scoffed at penny-ante tidbits like $2 + 2 = 4$, or nickel-and-dime fare like $5 + 5 = 10$. I wanted to go even further, to extrapolate new and fancier facts.

So, I reasoned, 24 being one less than 25, and 49 one less than 50, it must be that $24 + 24 = 49$. I divulged this wisdom to my teacher, like Prometheus delivering fire to humankind. "I love that thinking," she said. "But actually, you've taken away one from the first 25, and one from the second 25. So the answer isn't *one* less than 50. It's *two* less, which is 48. Still, nice reasoning!" She walked cheerily away.

I sat there, wounded and rebuffed, a king in a crumbling castle.

In our culture, addition facts are emblems of certainty. "2 + 2 = 4" is shorthand for "Some truths are undeniable." In George Orwell's *Nineteen Eighty-Four*, when the totalitarian government wants to break the hero's spirit, they torture him until he agrees that 2 + 2 = 5. Not just until he *says* it, but until he *believes* it. To Orwell's view, simple addition is the last refuge of truth, the hardest reality to deny—and thus, for aspiring tyrants, the final outpost of total control. Addition is the first operation we teach to children and the last operation that many feel comfortable with. It is a reminder of simpler times, a throwback to when things made sense.

Then again, go ask a boy whose eraser shavings and teardrops are blotting out 24 + 24 = 49. He will tell you that addition is not always so simple.

On first encounter, addition is a lot like stacking dishes. You put plates with plates, bowls with bowls, and cups with cups. The same principle applies to, say, 634 + 215: Stack hundreds with hundreds (600 + 200), tens with tens (30 + 10), and leftovers with leftovers (4 + 5). The total is 800 plus 40 plus 9, better known as 849.

But the dish analogy goes only so far. No stack of cups will ever yield a bowl. No number of bowls can ever be converted into a plate. Yet 10 ones make a ten, 10 tens make a hundred, and 10 hundreds make a thousand.

Hence, the big idea of addition, the fountain of its joys and challenges: *regrouping.*

For example, let's add 46 + 28. We stack the tens (40 + 20). We stack the ones (6 + 8). But that gives us enough ones for another full group of ten. So we create a new ten—in effect, turning a pile of cups into a single large bowl—and move it over to the tens pile.

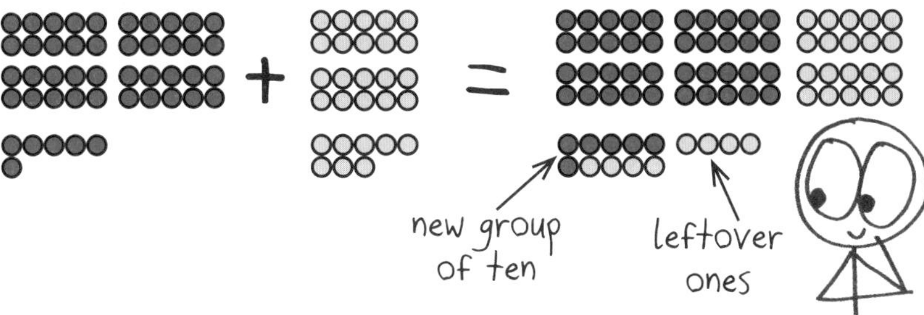

This method, known as *carrying*, is a "standard algorithm": a foolproof technique covered in every textbook, classroom, and YouTube explainer. Like all things standard, it is widespread, effective, and a bit overrated. Most math teachers (and I include myself) get less jazzed about standard algorithms than nonstandard ones.

What do I mean? Well, try plucking two units from 46 (so it falls to 44) and giving them to 28 (so it rises to 30). Just as politicians sometimes dodge a question by answering a different question that no one asked, regrouping lets us escape difficult operations (46 + 28) by transforming them into easier ones (44 + 30). Now our sum is a tidy 74, with no carrying required.

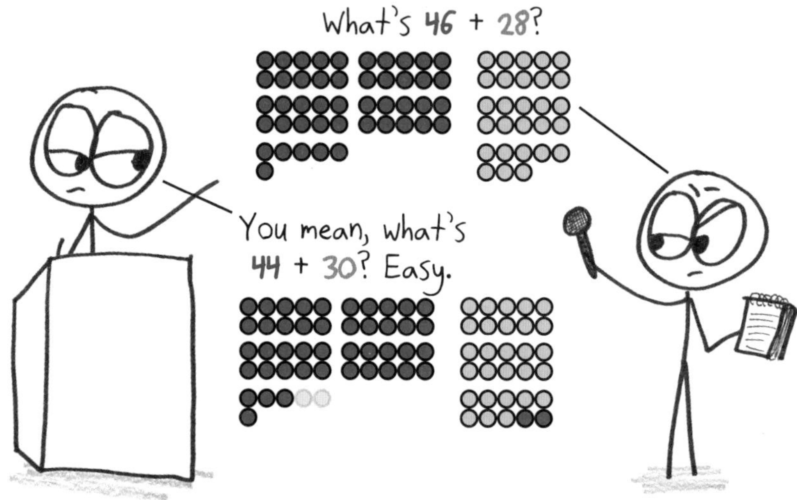

There's a tale the mathematician Carl Gauss loved to recount in his elder years: his childhood battle of wits with a scowling teacher named Büttner. Mr. B. told the class to sum the numbers from 1 to 100; each pupil was to place his chalkboard in the pile when finished. Mere moments later, seven-year-old Carl strode to the table and deposited his slate with a cry of "There she lies!" Büttner prepared to punish the "hasty young clerk" for his "frivolity." But, no, Carl had calculated true.

How did little Gauss make such swift work of that drudgery?

1 + 2 + 3 + 4 + 5 + 6 + 7 + 8 + 9 + 10
+ 11 + 12 + 13 + 14 + 15 + 16 + 17 + 18 + 19 + 20 +
21 + 22 + 23 + 24 + 25 + 26 + 27 + 28 + 29 + 30 +
31 + 32 + 33 + 34 + 35 + 36 + 37 + 38 + 39 + 40 +
41 + 42 + 43 + 44 + 45 + 46 + 47 + 48 + 49 + 50 +
51 + 52 + 53 + 54 + 55 + 56 + 57 + 58 + 59 + 60 +
61 + 62 + 63 + 64 + 65 + 66 + 67 + 68 + 69 + 70 +
71 + 72 + 73 + 74 + 75 + 76 + 77 + 78 + 79 + 80
+ 81 + 82 + 83 + 84 + 85 + 86 + 87 + 88 + 89 + 90
+ 91 + 92 + 93 + 94 + 95 + 96 + 97 + 98 + 99 + 100

Simple: he changed the question. Specifically, he married off the hundred numbers into 50 pairs: smallest with largest (1 + 100), second-smallest with second-largest (2 + 99), and so on, all the way to up 50th-smallest with 50th-largest (50 + 51).

Thus rearranged, each of the 50 pairs has a sum of 101. That yields 50 hundreds (5000) plus 50 ones (50), for a total of 5050.

Such is the magic of regrouping. It changes the question while preserving the answer.

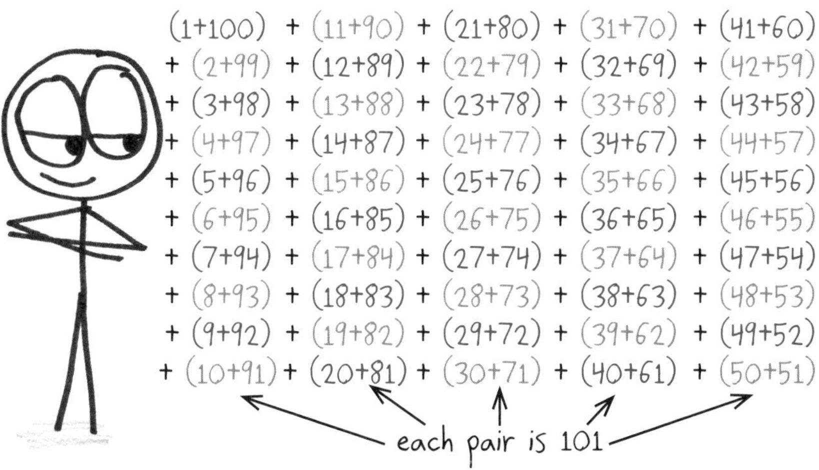

Orwell and the 2 + 2 = 4 crowd celebrate addition for its rigidity. But Gauss and the regroupers celebrate the opposite virtue: flexibility. When all that matters is the total, we can rearrange the parts however we please. We can shift items from one pile to another (as with 46 + 28). We can pair up smaller piles to make bigger piles (as young Carl did). And we can even arrange the piles into lovely shapes, bringing a dash of geometry to our arithmetic.

For example, what if instead of the first hundred numbers, Büttner had asked Gauss to sum only the odd ones?

1 + 3 + 5 + 7 + 9 + 11 + 13 + 15 + 17 + 19 +
21 + 23 + 25 + 27 + 29 + 31 + 33 + 35 +
37 + 39 + 41 + 43 + 45 + 47 + 49 + 51 + 53
+ 55 + 57 + 59 + 61 + 63 + 65 + 67 + 69 +
71 + 73 + 75 + 77 + 79 + 81 + 83 + 85 +
87 + 89 + 91 + 93 + 95 + 97 + 99

= ???

It's best to build up slowly. Starting with 1, we have a simple one-by-one square.

1 ● 1×1

Now, add in 3, organizing it as a nice L shape. That makes a two-by-two square.

1 + 3 2×2

Throw in 5, arranged as another L shape, and you've got a three-by-three square.

1 + 3 + 5 3×3

Next, adding in 7 gives a four-by-four square.

$$1 + 3 + 5 + 7 \quad 4 \times 4$$

We can keep going. The first five odd numbers make a five-by-five square. The first 10 odd numbers make a 10-by-10 square. Finally, the first 50 odd numbers—which is what we're trying to add—make a 50-by-50 square.

Hence, the sum is $50^2 = 50 \times 50 = 2500$.

Not long ago, after I recounted the tale of Gauss and Büttner in a conference talk, a professor approached me with a polite objection. I thought he might question the veracity—historians disagree on whether it really happened—but he had something else on his mind. "We tell the story as if Büttner were a villain," he said. "But he became a great advocate for Gauss—he found him a tutor and helped him into a life of mathematics." I like that version. Beneath Büttner's rigid surface was a teacher of warmth and flexibility, just as beneath the rigid certainty of $2 + 2 = 4$ is a flexible process of regrouping and rearranging.

I hope Orwell would agree: if a nation is a sum of people, then perhaps democracy is the process by which we regroup and rearrange ourselves, striving to turn difficult questions into ones we can answer together.

Subtraction

One evening when I was five years old, my sister, Jenna, gave me a worksheet of arithmetic problems. (Not the most inspiring lesson, but I can forgive a rookie teacher, especially one who's only eight.) I dutifully carried out the additions. But that left half the problems undone. These ones had an unfamiliar format: two numbers flanking not a + symbol, but a strange and meaningless horizontal line: –. What was I meant to do?

At last, I realized the trick. The vertical part of the + had been omitted, left as a kind of warm-up exercise for the reader. I went through drawing vertical lines, converting each – into a +, and then solved the now-sensible problems: $7 - 4$ became $7 + 4$, which was 11.

Jenna reacted as if I'd graffitied her pajamas. "It's not *addition*," she told me. "It's *subtraction*."

Thirty years later, I found myself reliving that moment in reverse. I mentioned this chapter to my wife, Taryn, who furrowed her brow and cautiously queried whether the chapter was truly necessary. "Do you want to pretend that subtraction is its own operation?" she asked. "Or will you explain that it's really just addition?"

Taryn, you see, is a mathematician. In the ways of her tribe, subtraction is mere shorthand for adding a negative. What you call 5 – 3, her people call 5 + –3. You don't begin with five apples and then give away three of them; you begin with five apples and then acquire three anti-apples.

To put it mildly, this view is counterintuitive. It seems to overlook the simple fact that putting two things together is not the same as taking one away from the other.

Say I bring $71 (seven $10 bills and a $1 bill) to the store and wind up spending $48 on mangoes and cream soda (a fair approximation of my weekly grocery outlay). I have clearly not gained any money. Rather, money has left my possession. If I want to know how much cash remains, addition is of no use; I must subtract the $48 from the $71.

Taking away the first $40 is easy; I'm left with $31.

Easy enough, after that, to take away another $1. I'm left with $30.

But how do I take away the final $7? I'll need to cash out one of my $10 bills into ten $1 bills.

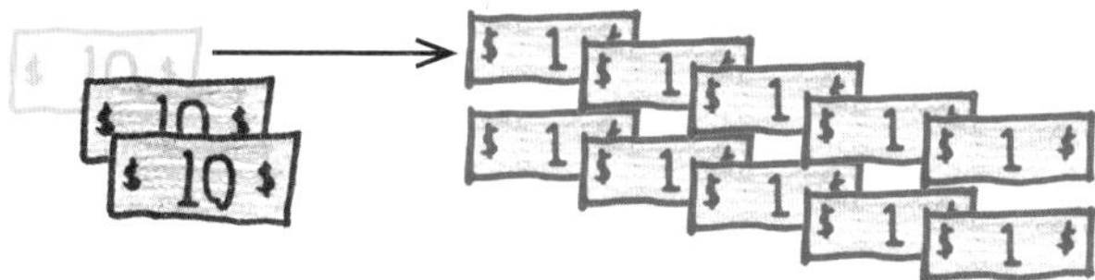

Then I can pay the last $7 I owe, leaving me with $23.

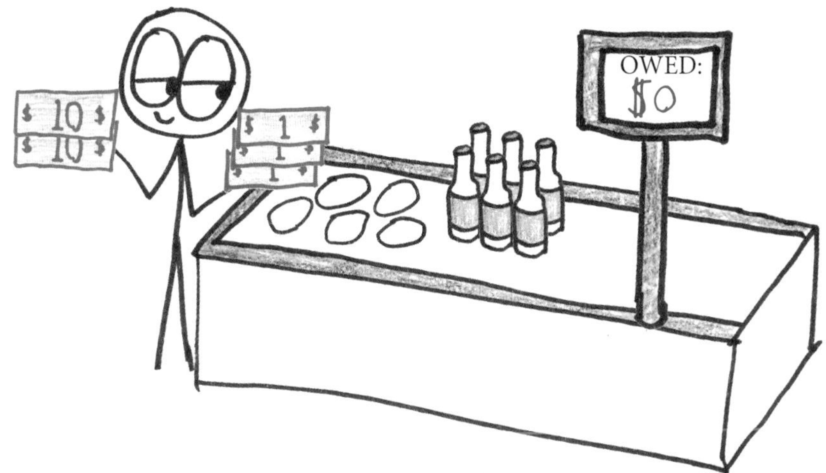

This is, more or less, the standard algorithm known as *borrowing*. Whereas addition forces us to turn a pile of small bills (say, ten $10 bills) into a big bill (one $100 bill), subtraction requires the opposite, converting one big bill (say, a ten) into a stack of small bills (10 ones). Arguably, the terms "carrying" and "borrowing" should be replaced with "combining small bills" and "breaking down big bills."

But whatever your preferred language, "addition" and "subtraction" aren't synonyms. They're antonyms. In mathematics, we call them *inverse operations*: each undoes the other. To add 3 and then subtract 3 is to land right back where you started: locking a door and then unlocking it.

But Taryn's dogma finds support in another use of subtraction: distances. For example, if I began the day 71 miles from home, and have driven 48 miles already, how much farther do I have to go?

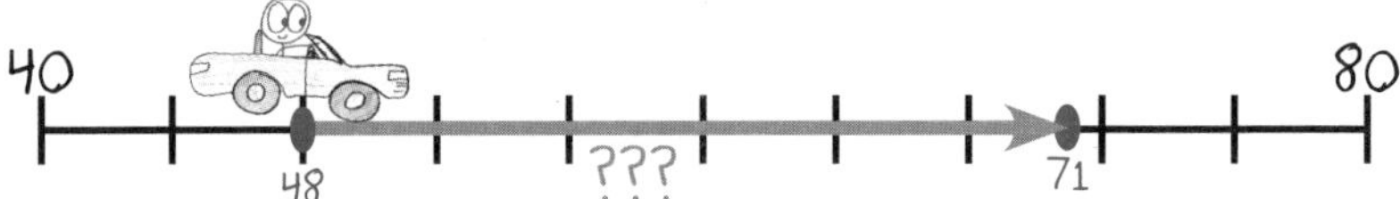

Well, after another 2 miles, I'll have driven a nice, round 50.

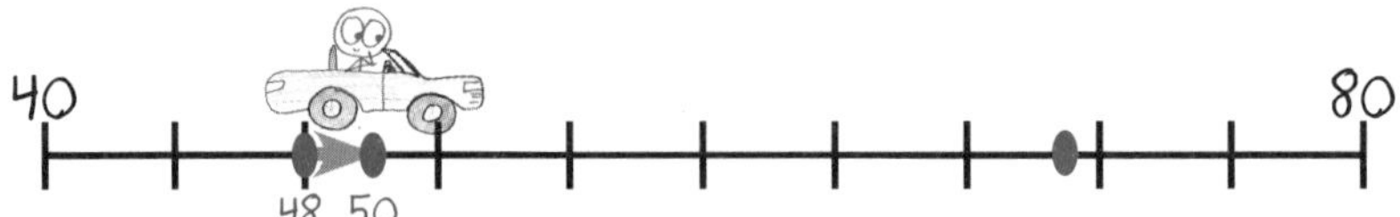

Then, after another 20 miles, I'll be up to 70.

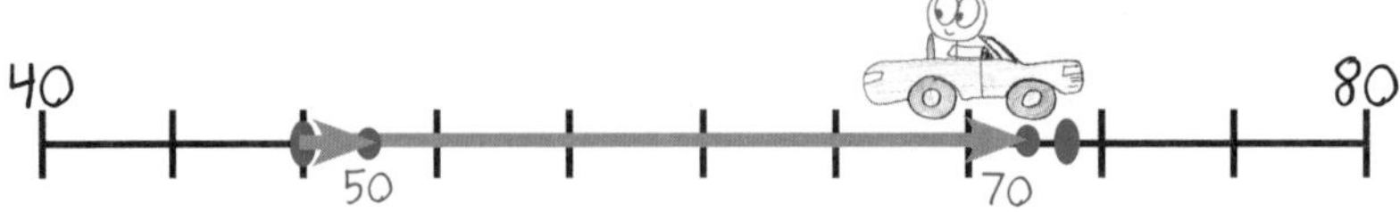

That leaves just 1 mile. So, overall, I need to travel 2 + 20 + 1, for a total of 23 miles.

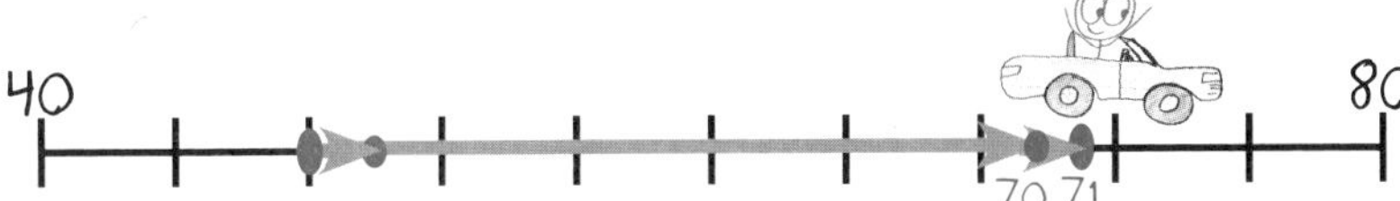

Suddenly, Taryn's view begins to gain a glimmer of plausibility. Just now, to solve a subtraction problem, I worked by successive additions: plus 2, plus 20, plus 1. This additive process gave the same result as the mango-and-soda expenditure, yet involved no taking away. Indeed, what would it mean to "take away" 48 miles from 71 miles? I'm only driving on the highway, not stealing a segment of it.

Taryn's "no such thing as subtraction" idea boils down to a simple proposal. Instead of seeing +3 and –3 as opposite processes using the same number, why not see them as the *same* process using *opposite* numbers?

This quirky theology has its benefits. First, it streamlines math, reducing two operations to one. And second, it helps to resolve some tricky ambiguities.

For example, in \$8 – \$3 – \$1, what is being subtracted from what? Evidently, I began with \$8, and I spent \$3—maybe on a coffee. But what happened next? Is the – \$1 another expenditure, say a muffin, to be subtracted from the original \$8? Or is it a reduction of the *earlier* expenditure, say a coupon for the coffee, to be subtracted from the \$3? The first interpretation lands on \$4 (which math teachers consider correct). The second lands on \$6 (which they do not).

Taryn's view offers a simple solution: rewrite the whole thing as $8 + -3 + -1$. No subtracting positives, just adding negatives. Whereas subtraction is fragile and fussy, addition is easygoing, giving the same result no matter what order of operations you choose.

So if Taryn is right after all, why is this chapter still here? Shouldn't I remove it? Or, rather, make it vanish from the book by adding its opposite?

In theory, yes, subtraction is reducible to addition. But you and I, unlike Taryn and her tribe, don't live in Theory Land. We live on the bustling, stinky streets of reality, where there are highways to traverse, mangoes to purchase, and other tasks that do not benefit from the mathematician's abstract worldview. Better to do as a wise eight-year-old once commanded, and embrace subtraction as a process all its own. As Niels Bohr once said, the opposite of a profound truth is another profound truth.

Multiplication

As October rolled around, the sixth graders looked ahead to a big exam, and my class began peppering me with questions about the prior year's version. Was it long? Was it difficult? Was it a living torment? I realized I hadn't taken it myself, and so, figuring I should try whatever I'm asking of my students, I breezed through it in about five minutes.

Perhaps I should have spent another five, because I did not score 100%. Instead, I calculated the wrong answer to 2573 × 389.

Multiplication is sometimes described as repeated addition. But to me, it's the operation of making rectangles. To multiply 8 × 4 is to find the area of an eight-by-four rectangle, or the number of items in an eight-by-four array.

These rectangles explain a great deal. For example, why doesn't order matter in multiplication? Why is 7 × 3 equal to 3 × 7? In terms of repeated addition, it's not obvious: Why should 7 + 7 + 7 give the same result as 3 + 3 + 3 + 3 + 3 + 3 + 3? But the rectangles explain it nicely: $a \times b$ and $b \times a$ are the same rectangle, just turned on its side. Thus, in mathematical jargon, multiplication is *commutative*.

Unfortunately, such rectangles aren't much help with 2573 × 389, unless you are a very patient counter of dots. For such tasks, we fall back on a standard algorithm: pushing the numbers around the page, in accordance with memorized rules. I knew the rules, of course; my downfall had been a clumsy error.

But a vain and stubborn part of me refused to accept defeat. So, not telling him why, I asked my colleague Ed—a sharp, intuitive mathematician, as well as the sanest person I know—to carry out the multiplication himself.

Ed botched it, too. I sighed with relief.

Then our friend Tom stopped by. I considered him, Ed, and myself to be the three musketeers of the math department, dashing young teachers of exceptional swordsmanship. But when Tom learned of the blunders Ed and I had made, I feared the musketeers were through: he gazed at us with contempt and pity, and sat down to show us how it was done.

Tom refused to believe he had erred until two different calculators refuted him.

Discovering my friends were as sloppy and careless as I am, I felt increasingly vindicated, and began to indulge my worst habit: openly philosophizing. "You see?" I said. "We shouldn't be asking this question at all. This isn't the interesting part of multiplication."

For example, I was more interested in *factorization*: rewriting a given number as the product of two others. Or, in geometric terms, arranging a set number of dots into a perfect rectangle.

Some intriguing numbers, such as 24, can be factorized in many ways.

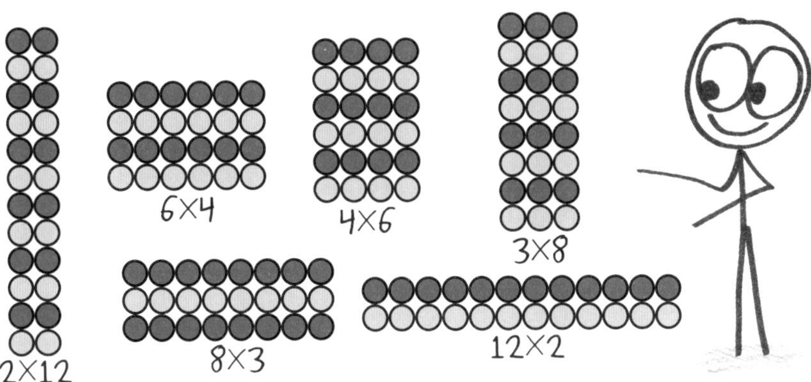

Even more intriguing are numbers we cannot factorize at all. No rectangle works for them (other than the boring rectangle of a single row). For these, I would propose the term "lumpy numbers," or perhaps "nonrectangular monstrosities," but mathematicians have settled on a loftier title, which I must admit has a nice ring to it: "prime numbers."

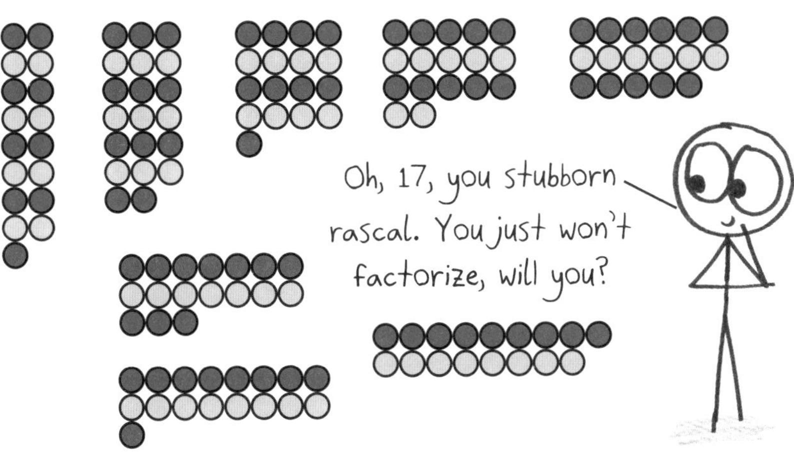

Isn't this business of factors and primes more interesting than 2573 × 389?

"Well, of course," said Ed, ever-patient with my flights of fancy. "But should 'interesting' be our only priority as teachers? Or should our students also—and I'm just spitballing here—be able to multiply two numbers together?"

I grudgingly admitted that he had a point.

Now the department guru arrived on the scene: Richard, a PhD mathematician and an intellectual hero of mine. We showed him 2573 × 389 and explained that none of us could manage it. "Ooh," he said, as if this were a juicy puzzle. "Are the factors in different bases? Are we doing modular arithmetic? Is it some kind of notational trick?" No, no, we said. Just multiply the numbers. That's what we're struggling with.

The joy drained from his eyes. "Pitiful," he said, and grabbed his pen. Moments later, the most vaunted math department in England's West Midlands was 0 for 4.

How exactly does one multiply? In her book *A Compendium of Mathematical Methods*, Jo Morgan lays out a baker's dozen of algorithms. They include the grid method, the lattice method, the stick method, and geographical variants ranging from ancient Egyptian to Russian peasant. In our failures, my colleagues and I had employed the traditional method known as *long multiplication*. "It is efficient and relatively straightforward," writes Jo, "though like other algorithmic approaches is often performed by pupils who do not understand why it works."

Long multiplication, like most methods, relies on the *distributive property*. This is the simple fact that one big rectangle can be broken into two small rectangles. For example, 17 × 6 can be broken into 10 × 6 plus 7 × 6. Makes sense; after all, 17 somethings is the same as 10 somethings plus seven somethings.

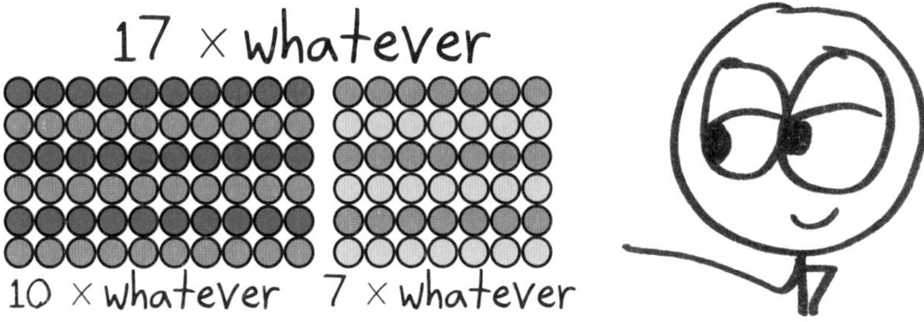

Long multiplication boils down to the distributive property, applied as many times as necessary to make the problem tractable. For example, to multiply 27 × 38, we first treat this as "38 somethings," and then break it into "30 somethings" plus "8 somethings."

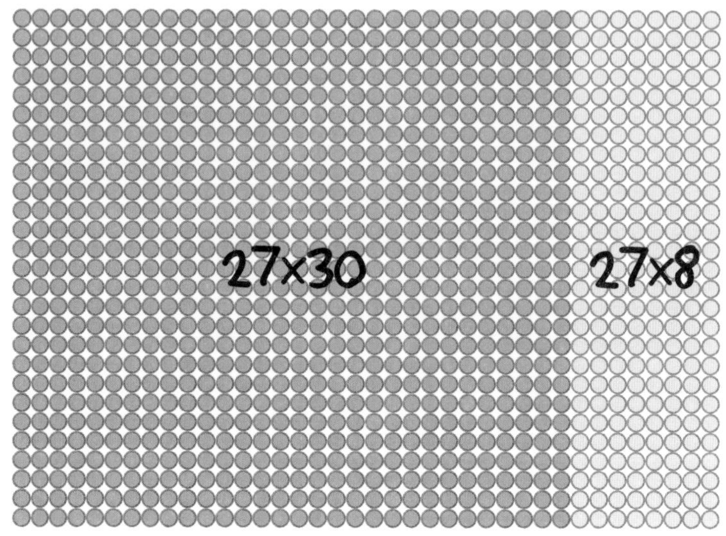

Then, since each of those products is "27 somethings," we break each one into "20 somethings" plus "7 somethings."

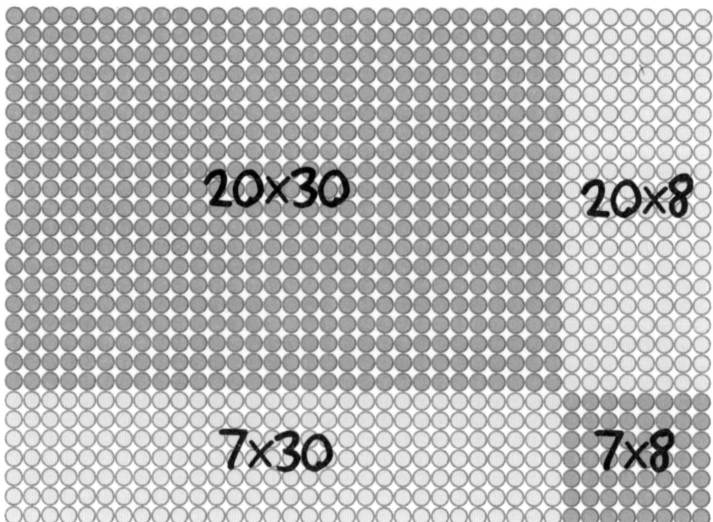

A single unmanageable multiplication thereby becomes four quite manageable ones. All that remains is to add them together.

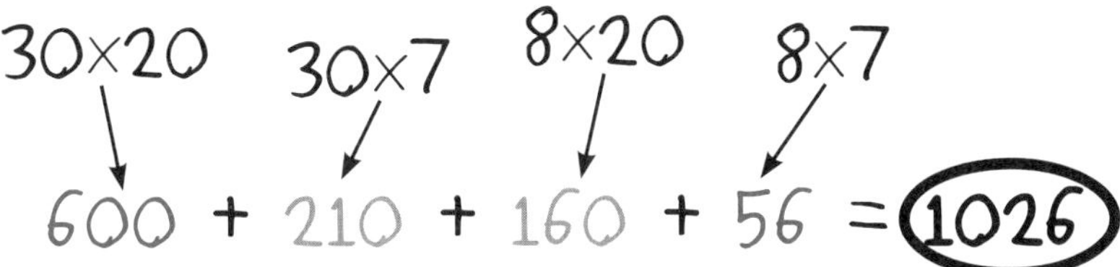

But if multiplication is that simple, why our mishaps and struggles?

Well, 2573 × 389 doesn't break into three or four multiplications. It breaks into 12. After that, to find their total, you must carry out 11 additions. Every step is a new opportunity to slip up. Someone who makes a careless error on just one in every 30 operations is more likely to get 2573 × 389 wrong than right.

On a typical day, I'm sure that all four of us would have managed it with ease. But on that day, something atypical was in the air.

I found Simon in the staff room. Of all the math teachers, he was the most careful and the most competitive, a man who'd commit hard fouls to win backyard soccer matches against his own nine-year-old. If anyone could execute 23 operations without a careless error, it was Simon.

Guess what happened next.

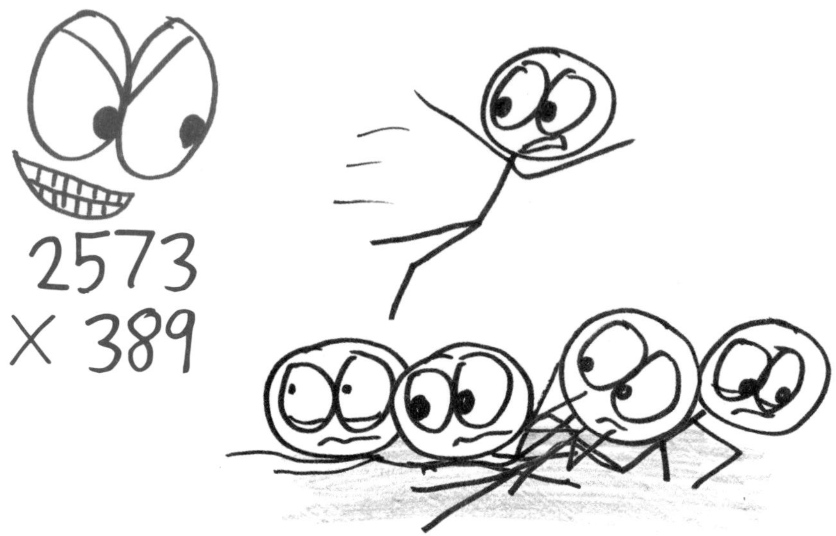

My point in telling this story—other than to drag my dear friends through the mud—is that the language of multiplication works on two levels. On a deep level, it is a language of rectangles, expressing abstract concepts like commutativity, primality, and the distributive property. To speak this level requires insight and understanding. But on another level, multiplication is a language for calculating products, a system for finding answers. To speak this level requires patience and precision—virtues that, at least on this day, we lacked.

Finally, we brought the problem to our department head, Neil. After recounting the long and winding tale of our failure, we asked him to try it himself.

"Oho, certainly not," he chuckled as he strode away.

We were still standing there, agreeing that Neil was the wisest among us, when along came Emily: a cheerful trainee Latin teacher who had last taken math at the age of 16. "I'll have a go!" she announced, and we all watched as she carried out 12 multiplications, along with 11 additions, to arrive at the correct answer of 1,000,897.

Division

One day in college, my education professor gave us the task of turning operations into stories. Shown a calculation, say 28 ÷ 4, you then had to embody it in a scenario. For example: "Four people share 28 cookies. How many does each person get?"

As a newly minted math major, I considered this lesson a slap in the face. *C'mon!* I thought, *I know about Lie groups. Or at least I took a course on Lie groups. Why would I need to practice fourth-grade arithmetic?*

"Now," said my professor, "here's a division problem that almost every math teacher in Japan gets right. But only a tiny fraction of teachers in the US can generate a story for it."

I sighed, anticipating another softball. Sixty people sharing 61 cookies? Three people sharing eight and a half cookies? Ah, what scintillating variety!

"What's a story," my professor said, "for 17 divided by ½?"

My eyes halted midroll.

I had always thought of division as the cookie-sharing operation. Cookie-sharing stories make sense when the final result is a fraction (17 cookies among two people), or even when the total being shared is a fraction (17½ cookies among two people). But what about a fractional number of people? What would it mean to share 17 cookies among ½ a person?

"How about this," I said. "Half of each person's share is 17 cookies. How many cookies in a person's whole share?"

My professor rotated his hand from side to side, in the universal gesture for *Eh, kind of, but not really.* "Wouldn't that be a better story for 17 × 2?" he said.

Ha! Now I had him. "Yes, of course," I said, "but that's because multiplication and division are inverse processes. Dividing by ½ is the same as multiplying by 2. So any story that works for $17 \div \frac{1}{2}$ will necessarily work for 17 × 2, and vice versa."

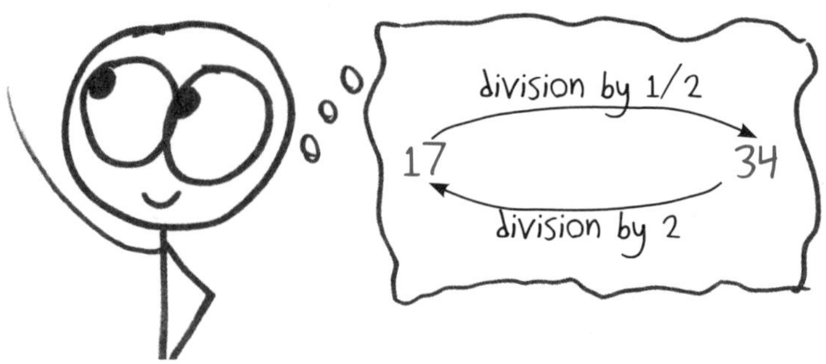

"In theory, I suppose," he said. "But if I had two boxes, each with 17 books, would you really model that as $17 \div \frac{1}{2}$?"

He had a point. I hate it when people have points.

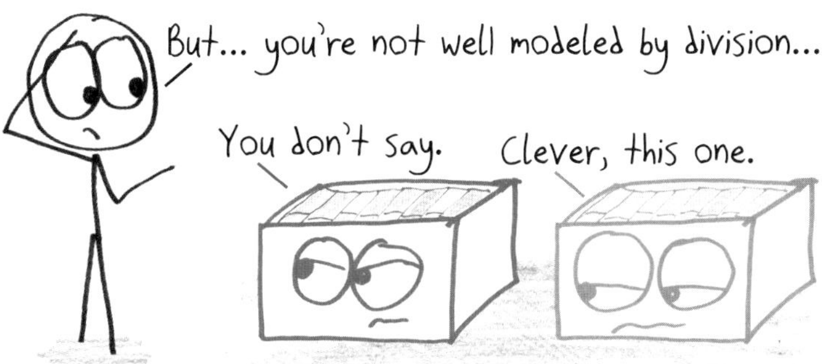

By one view—the view I held at age 21—the beauty of mathematical language lies in its unreality. Math lets us escape this world of crumbs and mud for a realm of rigorous abstractions. The purer the logic—that is, the further from physical reality—the deeper the truth. "As far as laws of mathematics refer to reality, they are not certain," said Einstein, "and as far as they are certain, they do not refer to reality."

I considered the physical world a small price to pay for certain truth. Under my view, $17 \div \frac{1}{2}$ and 17×2 were indistinguishable. Dividing by a number is the same as multiplying by its reciprocal, always and forever.

Then again, is this true fluency? Or was I like a renowned chef who can't make a peanut butter sandwich? What good is dividing 17 by ½ if you can't describe an occasion when you'd need to? "Dividing one number by another is mere computation," the mathematician Jordan Ellenberg once wrote. "Figuring out *what* you should divide by *what* is mathematics."

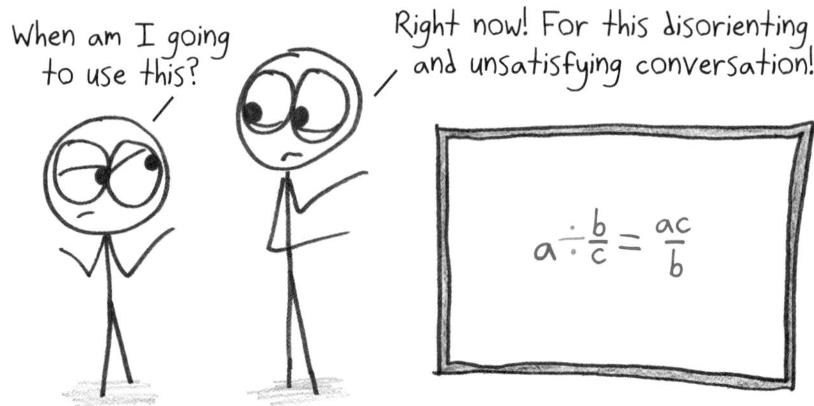

For a better understanding of division, we need to go back to multiplication. Any multiplicative act lends itself to two subtly different interpretations, depending on which number gives the size of the groups. When you say 2 × 5, do you mean two groups of five? Or five groups of two?

They're two distinct pictures: a pair of quintets vs. a quintet of pairs.

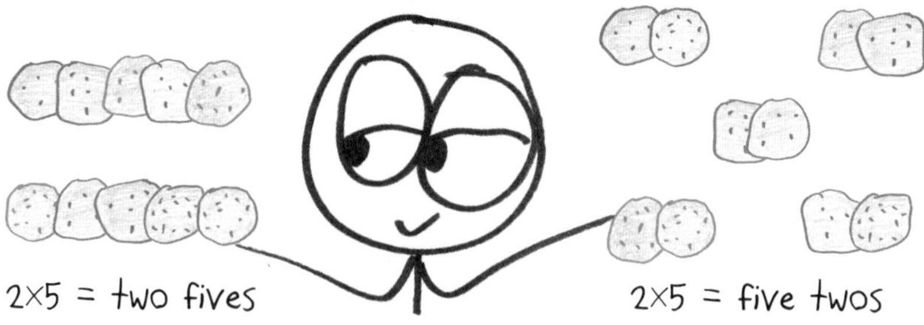

Now, because division is the inverse of multiplication, it permits two corresponding interpretations. When you say 10 ÷ 2, are you breaking 10 items into two groups? This is *cookie-sharing division*, or as education professors call it, *partitive division*.

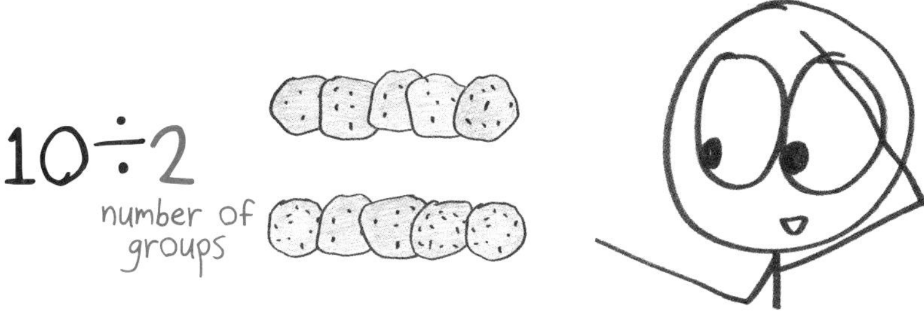

But here's the twist I was missing: Are you instead breaking 10 items into *groups of two items each*? I call this *bucket-filling division*: Given 10 gallons of water, how many 2-gallon buckets can you fill? The education experts call it *quotative division*.

For whatever reason (capitalism?), we in the US default to cookie-sharing. But to speak division fluently, we need both.

Take the famous truth that "you can't divide by zero." In cookie-sharing terms, 9 ÷ 0 is gibberish: it means sharing nine cookies among zero people, which is not a coherent question. But in bucket-filling terms, 9 ÷ 0 asks: Given 9 gallons of water, how many zero-gallon buckets can you fill? This question makes sense, but yields no definite answer, because you'll run out of buckets long before you run out of water. Hence, dividing by zero is an impossible operation (though it gestures in the direction of "unlimited" or "infinite").

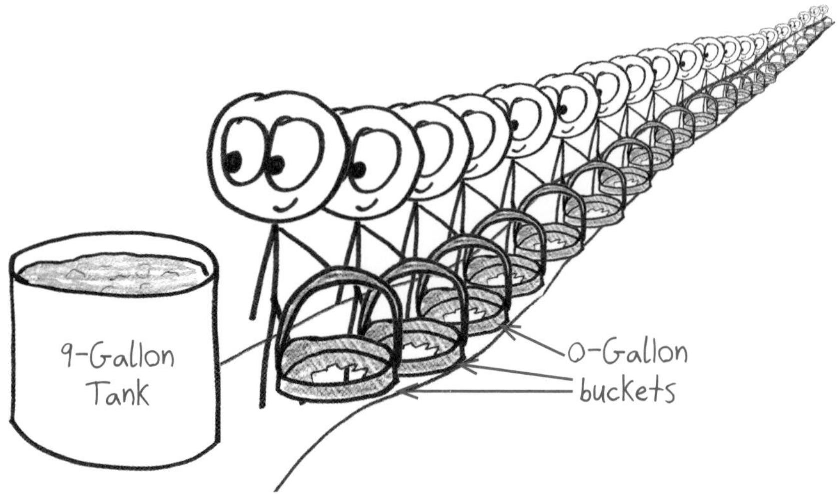

Though I speak English fluently, simple truths can stump and startle me. For example, isn't it weird that we have a word meaning "provide with food" ("feed") but no equivalent for "provide with drink"? (Long ago, this word was "drench.") To love a language is to open yourself to perpetual surprises.

And surprise is just what I felt when, after an agonizing two minutes, my professor stated a simple story for dividing 17 by ½. "How many half-dollars," he asked, "are there in $17?"

How had I missed that?

I was so busy thinking about
cookies I forgot you existed.
sigh
I get that a lot.

I sometimes envision mathematics as a tower. It leads from the earthy crust of everyday experience (piles of cookies; buckets of water; half-dollar coins) up to the thin atmosphere of abstract concepts (Lie groups, whatever those are). There's pleasure and power in climbing to the upper floors. But there's an equal pleasure—and a different kind of power—in descending to the bottom. Down there, you can touch the foundations, poke at the joints where math attaches to the world, and fill your half-gallon bucket with a new kind of insight.

Squaring and Cubing

I try not to be surprised by what people don't know. Never heard of Canadians? That's fair; they're a softspoken bunch. Didn't realize beef comes from cows? Hard to blame you; most cows are unaware themselves. No judgment—that's my philosophy. The world is wide, the seas are deep, and not even the greatest trivia champion knows every triviality.

Still, I can't believe how many people it shocks that "squaring" refers to actual squares.

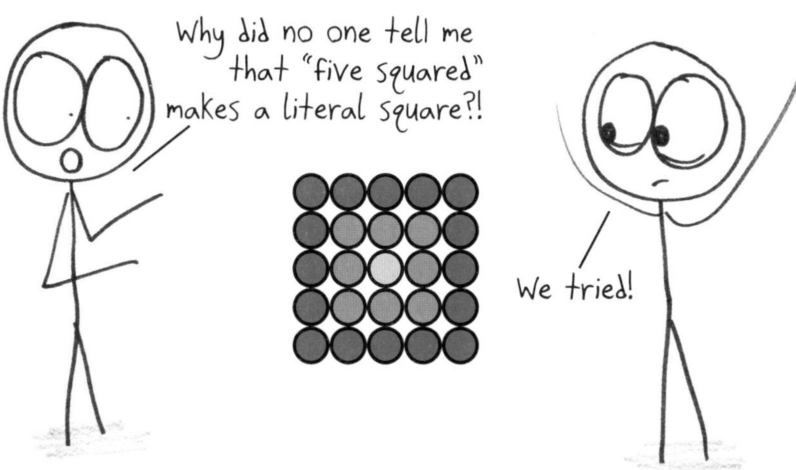

The symbol 5^2 and the pronunciation "five squared" mean the same thing. But they have clashing etymologies, a bit like the English-language distinction of "beef" vs. "cow." Back in medieval England, only the French-speaking elites could afford to dine on meat. Hence, the word "beef," from the Old French *boef*. Meanwhile, poor Anglo-Saxons worked the fields and tended to the animals. Hence, the word "cow," from the Old English *cū*. One etymology for the fancy food, and another for the lowly livestock—even though they're the same creature.

So it is in math. The word "squaring" comes from the earthy language of geometry, whereas the lofty little ^{2} comes from the airy language of algebra.

Squaring is one creature with two names.

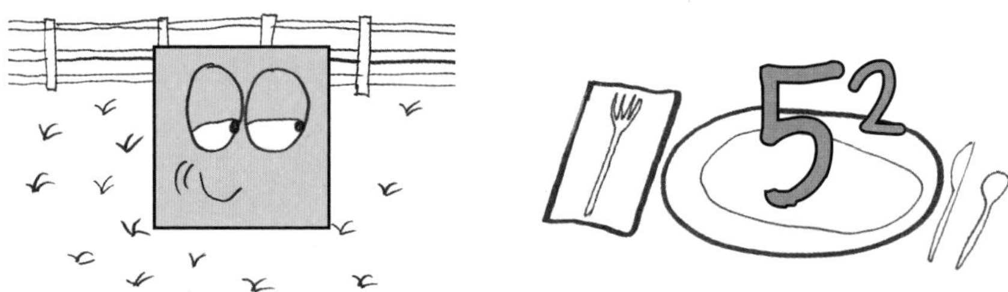

First, the geometry. As you may know, a square is a special kind of rectangle, with equal sides. Thus, squaring is a special kind of multiplying, with equal factors.

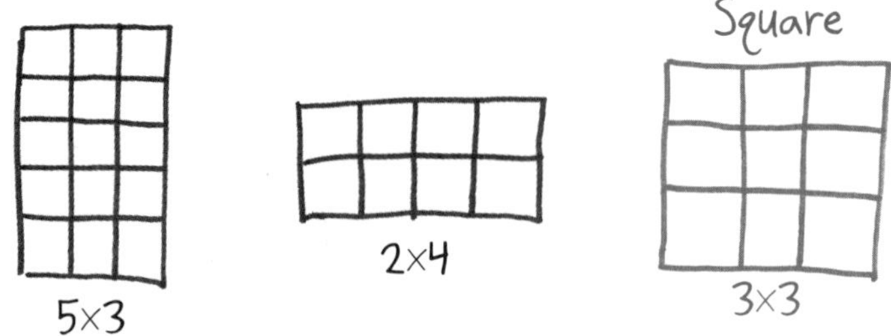

The same goes for "cubing." Multiplying three numbers forms a stack of rectangles known as a prism. When all three numbers are the same, that prism is a cube.

In the geometric worldview, a number is never just a number, but always some kind of measurement. One number gives a length. Two numbers (length and width) give an area. Three numbers (length, width, and depth) give a volume. For centuries, mathematicians would write a number, its square, and its cube using three totally different notations: something like l, q, and c. Distinct names for distinct geometries.

As for four numbers—well, that was a touchy subject. Length, width, depth, and...what, exactly? There is no obvious fourth dimension. So back in Euclid's day, scholars would bend over backward to avoid multiplying four numbers together.

But as time passed, the geometric worldview faded. Mathematicians came to see numbers, their squares, and their cubes as all belonging to a single family: "repeated multiplication." Hence, A's square became A^2 (short for $A \times A$), the cube became A^3 (short for $A \times A \times A$), and even the anomalous $A \times A \times A \times A$ was welcomed into polite society under the name A^4.

The new regime lets us state sweeping generalities in succinct and memorable form. It's great for swift computations. But it also has drawbacks.

Namely, the persistent, mistaken belief that $(a + b)^2$ equals $a^2 + b^2$.

We teachers fight this fallacy as best we can. We give examples. We present proofs. We scream into the void. None of it makes much difference. The error is too tempting; the algebraic language of exponents practically invites it.

By contrast, the geometric language of squares easily dispels it. Visit the cattle pasture, and you'll swiftly see that a 13-by-13 square encloses more area than a 10-by-10 square plus a three-by-three square. Geometrically, $(a + b)^2$ is not made of two smaller squares.

It's made of two smaller squares *plus two rectangles*.

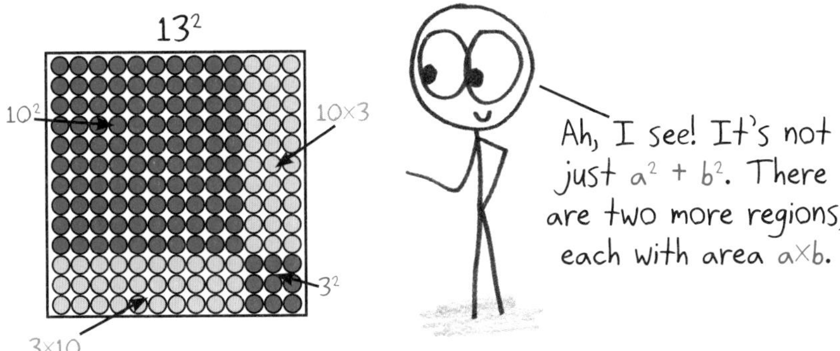

13²
10²
10×3
3²
3×10
Ah, I see! It's not just a² + b². There are two more regions, each with area a×b.

"Squaring" and x^2 are two names for the same idea. One is geometric, and one algebraic; one concrete, one abstract; one from the farmland, one from the banquet hall. Each knows something that eludes the other. And that's as it should be: the world is wide, the seas are deep, and not even the greatest notation has its finger on every truth.

Roots

One afternoon, Ashley dropped by with a math question.

She had recently arrived as a new 12th grader at the intense charter school where I taught. Whereas her classmates had spent years in the crucible, building up calluses to low grades and long homework, for Ashley the breakneck pace was novel—and a bit frightening. "Well," I said, pointing to the problem, "what does 'to the $^7\!/_2$ power' mean?" She lowered her head and apologized. A discouraging start.

Answering a student's math question often means digging into the past. You become a forensic educator, scraping through layers, seeking an old skill that they never quite acquired.

"Okay," I said. "What exactly is a square root?" She hesitated, and I knew that we had found our missing artifact. Ashley knew how to square a number, knew that 5 squared means 5 times 5, which is 25. But to reverse this process—to ask "Which number, when squared, gives 25?"—was a step she hadn't yet taken, a language she couldn't yet speak.

Easy to see why. Square roots are a doozy, the mathematical equivalent of an irregular verb.

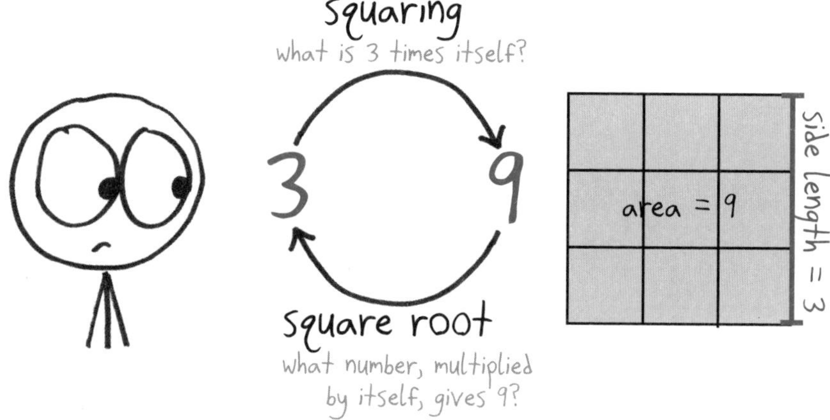

Roots go by many names. In the US, we call them *radicals*. In the UK, they call them *surds*. But the real trouble with roots, the thing that makes them so vexing, is neither their revolutionary character nor their existential absurdity.

The trouble is that roots are nasty, nasty numbers.

Ever tried calculating a square root by hand? It's a filthy business. The moment calculators became widely available, we jettisoned the task from the curriculum, for much the same reason that we stopped sending children into coal mines.

Not all roots are so dire, of course. Some, like $\sqrt{4}$ and $\sqrt{9}$, come out as nice, whole numbers. But the roots between those are irrational, expressible only as endless decimals. What exactly is $\sqrt{7}$? A calculator can give a rough figure (2.646-ish), but in lieu of infinite paper and time, the only precise way to write the square root of 7 is as some variation on the phrase "the number that, when squared, gives you 7."

Square roots can seem like a badly designed language. But the truth is even more troubling: they're a well-designed language for a badly designed reality. Go take a diagonal shortcut across a one-by-one square; you just walked $\sqrt{2}$ units. The shortest path across the simplest shape turns out to be an irrational length, a number we cannot name except by describing the operation that produces it: $\sqrt{2}$.

Roots thereby scramble the whole distinction between nouns and verbs. We write an operation, yet we treat it like a number.

Radical and absurd, indeed.

"Okay," I said to Ashley, trying to arrange my thoughts. "Here, I think you'll like this." To be honest, I didn't know whether she would, but I was confident she'd grasp the concept swiftly. I wrote a few symbols on the whiteboard and pivoted back to ask a question.

Ashley wasn't looking at the board. She was crying. "I'm sorry," she said. "We can keep going in a second. It's just…this has always been so hard for me…"

Her voice trailed off, but I knew the story. Ashley had spent years battling symbols she did not understand. She suffered the daily disappointment of a language that did not speak to her. I stood there: helpless, unhelpful, with no idea what to say.

A few weeks later, Ashley left the school. We wished her the best. I don't know if she wound up graduating.

I like to believe that our conversation gave Ashley a brief catharsis, that, if only for a fraction of an hour, she felt as smart as she is. I say I *like* to believe this, but really I *need* to believe it, because I know that in my perpetual state of haste and hurry, I have done the opposite for so many students: that by sprinting through algebraic steps on the board, by letting frustration or impatience creep into my voice, by passing back failing quizzes without ceremony or sympathy—that in all these tiny, terrible ways, I've made them feel stupid.

I saw Ashley one time a few months after she left our school, from the window of a bus. We both smiled and waved, and I managed not to cry until I got home.

Exponents

Here's a question I often wrestle with: Why don't people have a better understanding of exponential growth?

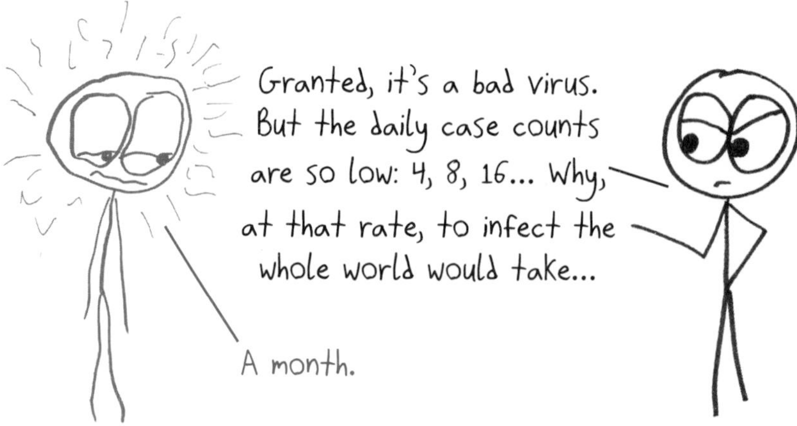

Granted, it's a bad virus. But the daily case counts are so low: 4, 8, 16... Why, at that rate, to infect the whole world would take...
A month.

Is exponential growth an *unnatural* idea? No, it's a pattern found throughout nature. Populations of all kinds, from bacteria to rabbits to humans, tend to grow exponentially.

Is exponential growth a *complicated* idea? Not really. Whereas linear growth occurs when you *add* a certain amount every step (2, 4, 6, 8, 10...), exponential growth occurs when you *multiply* by a certain amount every step (2, 4, 8, 16, 32...).

Is exponential growth just too *unfamiliar* an idea? Again, no. If you've seen the rise of a social media network, a pandemic, or even a viral video, then you've seen exponential growth firsthand.

So why all the trouble? One cause, I fear, is language. This essential concept arrives clothed in incomprehensible symbols. I refer to the vaunted and dreaded "exponent laws."

RULE	EXAMPLE
$x^0 = 1$	$2^0 = 1$
$x^{-n} = \frac{1}{x^n}$	$2^{-3} = \frac{1}{2^3} = \frac{1}{8}$
$x^{\frac{1}{n}} = \sqrt[n]{x}$	$2^{\frac{1}{3}} = \sqrt[3]{2}$ (roughly 1.26)

For years, we tell students that exponents are repeated multiplication: for example, 2^3 means $2 \times 2 \times 2$. Then, in an Orwellian reversal, we tell them that exponents are nothing of the sort. Some exponents are reciprocals, some exponents are roots, and some exponents, by royal fiat, are always equal to 1. It's as if the government redefined "theft" to include not just "taking someone else's stuff," but also "driving an ugly car" and "public sneezing." How could any citizen feel safe when legal terms are prone to arbitrary redefinition?

Yet these exponent rules are not as capricious as they seem. In fact, they are inevitable. If 2^3 is repeated multiplication, then these laws are the natural extensions. Together, they form a system of interlocking parts.

The negative exponent, as in 2^{-3}, is repeated *anti*-multiplication.

The fractional exponent, as in $2^{\frac{1}{3}}$, is *partial* multiplication.

And the zero exponent, as in 2^0, is *non*-multiplication.

To see what I mean, let's write the powers of 2 along a number line. These numbers are nice, but as yet, they don't mean anything, so let's also imagine a bacterial population, doubling every hour. It begins as a blob. After one hour, the blob has doubled in size. After two hours, it has quadrupled in size. After three hours, it is eight times its original size. And so on.

We thus tether our *exponents* to a case of *exponential growth*.

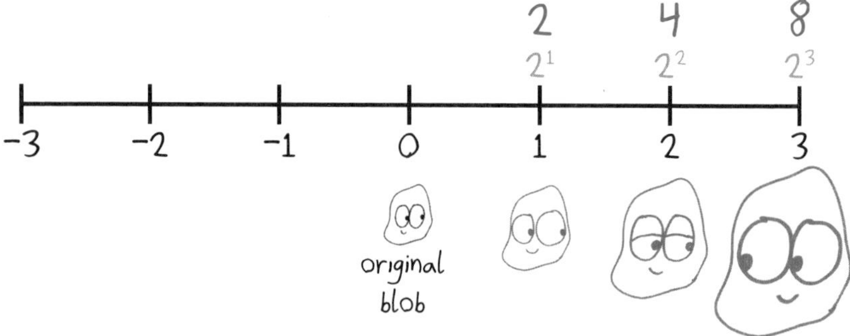

First, what about time zero? That was the moment we began measuring, so the blob was simply its original size: one blob's worth. It's a case of *non*-multiplication.

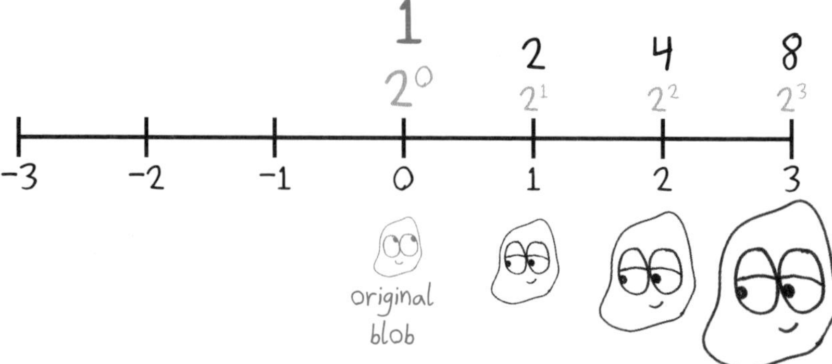

Every time we run the video forward an hour, the population doubles. So when we run the video *backward* an hour, the population will be cut in half. Keep running it backward, and instead of *multiplying* repeatedly, we will be *dividing* repeatedly.

Thus, at time –1 (one hour before we started), the blob must have been ½ the original size. At time –2 (two hours before we started), it must have been ¼ the original size. And so on.

This is repeated division. Repeated *anti*-multiplication.

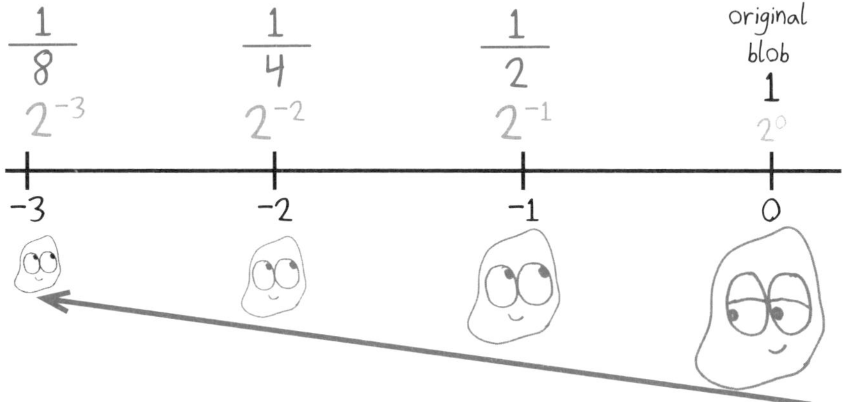

Finally, what about *between* the chimes of the clock? How big was the blob not after zero hours or one hour, but after *half* of an hour?

Be careful. When we go from 1 to 2, the halfway point is not, as you might guess, 1.5. That's *additive* thinking: 1 + 0.5 + 0.5 does indeed equal 2. But exponential growth is not additive; it's *multiplicative*. We don't want 1 + *something* + *something* to equal 2; we want 1 × *something* × *something* to equal 2.

Luckily, we have a name for this something: $\sqrt{2}$. After half of an hour, the bacterial population is $\sqrt{2}$ (roughly 1.41) times the size of the original blob.

Call it *partial* multiplication.

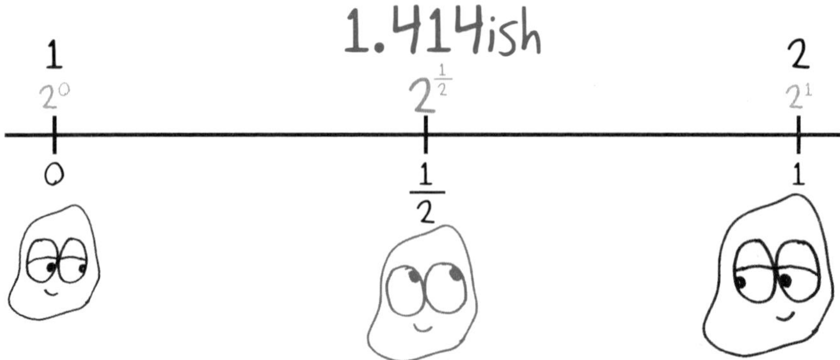

With such reasoning, we can find the size of the bacterial population at any time: positive hours, negative hours, the zero hour, even fractional hours. Lo and behold, the sizes correspond precisely to those troubling "exponent laws."

The third power is repeated multiplication.

The zeroth power is non-multiplication.

The negative third power is repeated division.

And the one-third power is partial multiplication.

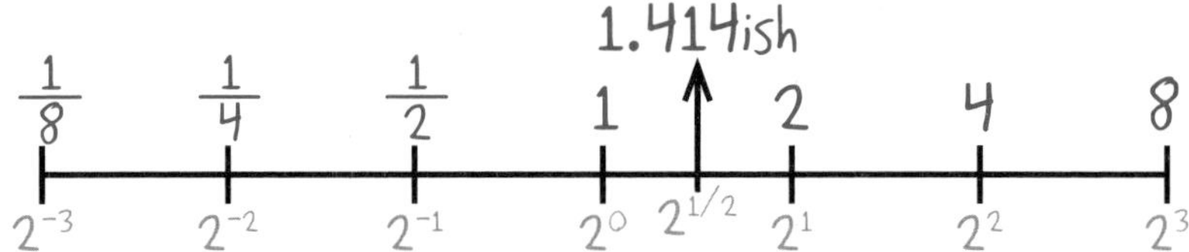

An old chestnut holds that poetry is "the art of giving different names to the same thing." The mathematician Henri Poincaré once replied that "mathematics is the art of giving the same name to different things."

Exponents offer a gorgeous case study. We take four different things and give them a single name. Repeated multiplication ($2 \times 2 \times 2$), repeated division ($\frac{1}{2 \times 2 \times 2}$), roots of various degrees ($\sqrt{2}$), and the plain old number 1 all become unified under the concept of an exponent. What makes this an art, rather than a hallucination, is that it works. These four seemingly disconnected operations are just phases in the life of exponential growth.

Not that any of this makes exponential growth an "easy" idea. Consider that the species *Homo sapiens* began with a little group of perhaps 10,000 people. It took a few thousand generations to reach a billion. After that, to reach 2 billion took another four or five. Not four or five *thousand* generations. *Four or five* generations. Hitting 3 billion took one more generation, and after that, 4 billion took only a decade. Exponential growth may not be unnatural, complicated, or unfamiliar. But it sure is *weird*.

Logarithms

On an island to the north of Great Britain, in a cozy bookstore in the town of Stromness, I bought a copy of *The Disappearing Dictionary*. In that book, linguist David Crystal documents words from various dialects of English, words that are full of local flavor but are in danger of fading from use: words like "dabberlick" (a tall, skinny person), "rumgumption" (common sense; shrewdness), and "squinch" (a narrow crack in the wall or floor). My personal favorite is "logaram," which I love not just for its meaning ("nonsense; rubbish; a long, embellished story") but for its etymology.

"Logaram" is derived from the ultimate form of balderdash: the logarithm.

The word "logarithm" was coined a few centuries ago, a few hundred miles to the south of Stromness. It is a portmanteau of *logos* (reason) and *arithmos* (number); literally, "logical number." Evidently, not everyone shared that favorable assessment.

In any case, logarithms were born for a definite purpose. They transform hard calculations into easy ones.

Back in the days before calculators, to carry out a long multiplication or division was a slow undertaking. (Witness my struggles in the chapter on multiplication.) But addition and subtraction? Those weren't so bad. Hence, the logarithm: instead of multiplying two numbers, you'd look up their logarithms in a big book, add the logarithms, and then convert the result back into an ordinary number. Logarithms thereby translate multiplication into addition.

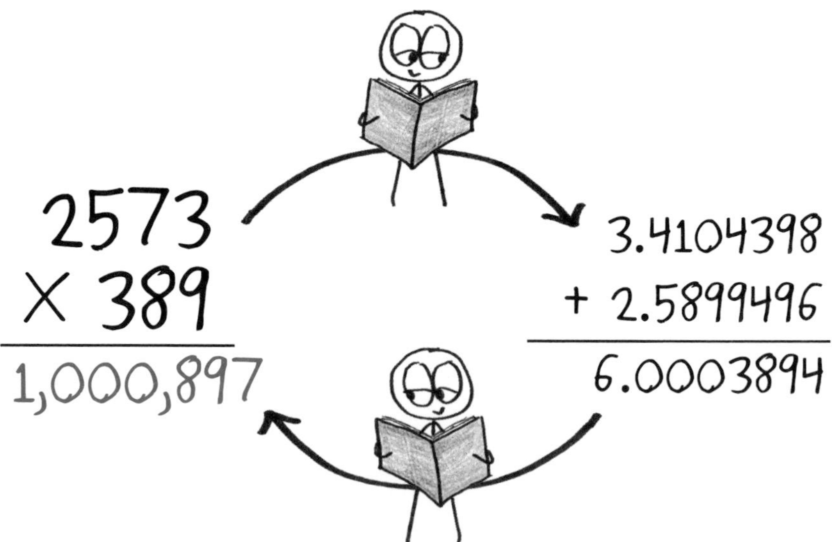

The effect is a kind of shrinkage. Logarithms (or logs for short) shrink multiplication into addition, division into subtraction, squaring into doubling, square roots into halving, and huge numbers into tiny ones.

In short, a logarithm is a mathematical shrink ray.

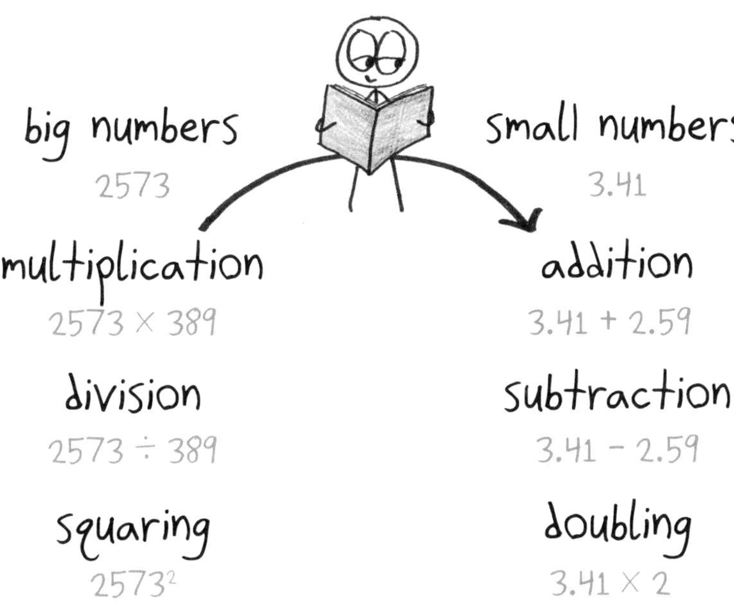

For a while, every scholar needed a book of logarithms by their side. "A Scottish baron has appeared on the scene...," the astronomer Johannes Kepler wrote to a friend, "who has done an excellent thing, transforming all multiplication and division into addition and subtraction." In 1823, the mathematician Charles Babbage promised the Exchequer of England that he could provide "logarithmic tables as cheap as potatoes." The prospect evidently delighted them.

Today, while potatoes still have currency, such tables are long obsolete. To us, the logarithm is no longer a labor-saving tool.

Instead, it is an operation. Specifically, the logarithm is the inverse of the exponential. An exponential spans orders of magnitude; a logarithm collapses them.

Take earthquakes. The largest one ever recorded (on May 22, 1960, in Chile) released almost a billion times more energy than the harmless little quakes that used to rattle my apartment in Oakland, California. How to handle such disparate sizes? With the logarithmic scale known as *moment magnitude*, which shrinks an exponential jump (×1000) to a linear jump (+2).

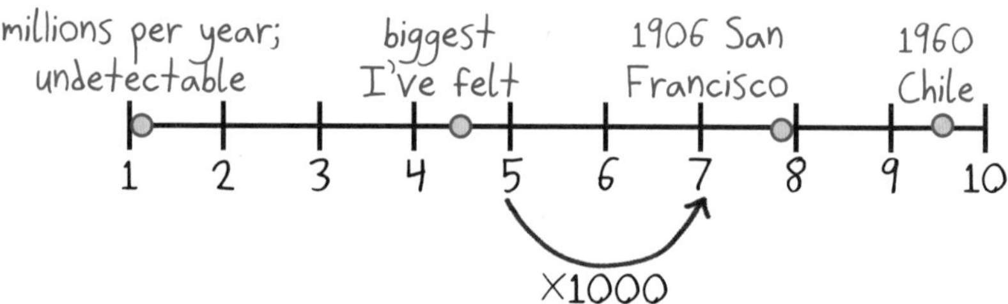

The same goes for sound waves. Thunder is as intense as a billion whispering voices, but the logarithmic scale of *decibels* (dB) can collapse these vast differences down to manageable ones. A multiplicative step (×10) becomes an additive step (+10).

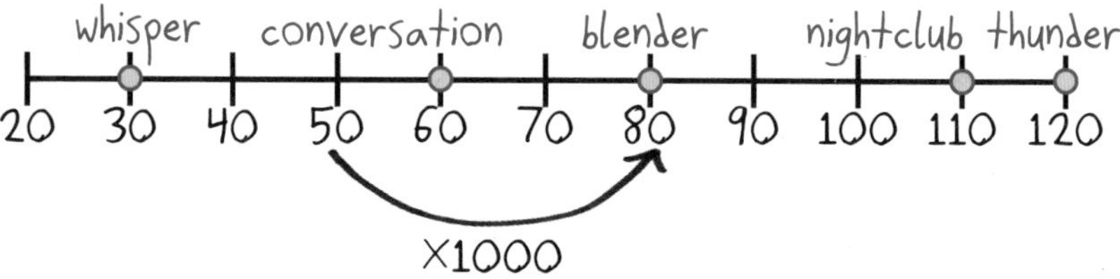

All across science, from the acidity of chemicals to the brightness of stars, we must translate multiplicative scales into additive ones. Logarithms are our ×-to-+ dictionary.

And a mighty useful dictionary they have been. Beyond science, logarithms spent centuries at the heart of basic work in engineering and navigation; as the science writer James Gleick put it, "Logarithms saved ships." But the logarithm paved the way for better calculating tools; it sealed its own obsolescence. The ×-to-+ dictionary was, in fact, a disappearing dictionary.

Or so one might have guessed. Though born for easing calculation, the logarithm proved useful far beyond it. In this way, the logarithm is a microcosm of the whole language of mathematics: born for narrow, practical purposes, it grows into a kind of deep and expansive literature—a literature that still makes space for the occasional logaram.

Grouping

I once stumbled across a cleverly compiled list of ambiguous news headlines. Three of them managed to lodge themselves permanently in my memory, each one bringing vivid, terrible images to mind:

Contrary to first impressions, nothing terribly exciting is afoot. The milk drinkers are *consuming* powder. The kids are *preparing* healthy snacks. And the *complaints*, not the refs, are becoming hideous. Each headline can be read two ways, but our knowledge of the world makes obvious which reading is correct.

Every language is prone to ambiguity, to what you say being taken in a way you didn't mean. But ambiguities pose special trouble for mathematicians, because theirs is a language of pure grammar. Looking at an equation, there are no context clues to help us.

For example, which operation comes first here: addition or multiplication?

$$2 + 3 \times 4$$

If we add first, we'll get 5×4, which is 20. If we multiply first, we'll get $2 + 12$, which is 14. So which one is right? Are the milk drinkers ingesting powder or becoming it?

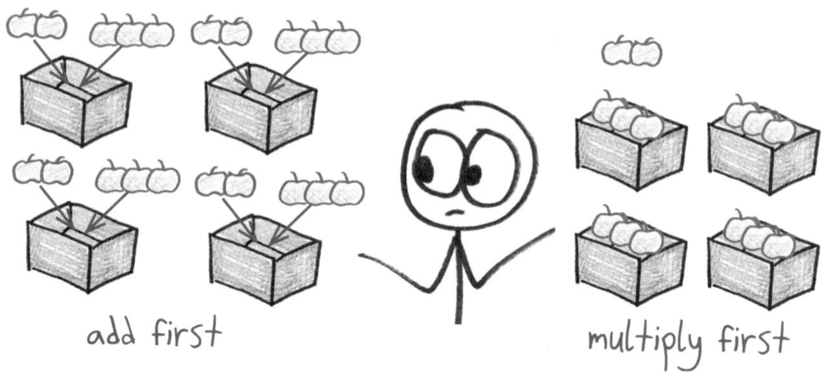

Here's another:

$$2 \times 3^2$$

Multiply first, and you'll get 6^2, which is 36. Square first, and you get 2×9, which is 18. Which reading is intended? Are the kids the chefs or the cuisine?

We need a consensus on how to resolve these ambiguities, an agreed-upon system for which operations to carry out first. Lucky for us, mathematicians have converged on a simple rule.

More powerful operations take priority.

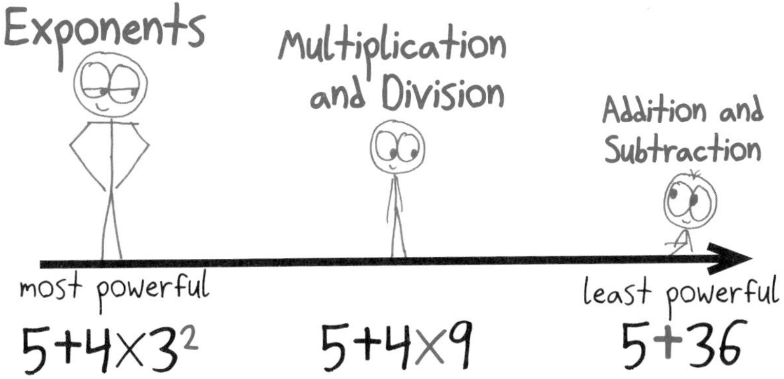

For example, multiplication is more powerful than addition. (Indeed, it's a form of repeated addition.) So in 2 + 3 × 4, we multiply first and add second. The result is 14.

Meanwhile, exponentiation is more powerful than multiplication. (Indeed, it's a form of repeated multiplication.) So in 2×3^2, we square first, and multiply second. The result is 18.

What about 8 – 2 – 5 + 4? Addition and subtraction are equally powerful. (They're inverses of one another.) So as a tiebreaker, we simply operate from left to right: 6 – 5 + 4, and thus 1 + 4, and thus 5.

Ambiguity resolved. But there's a problem: What if we want the *less* powerful operation to come first? Is there no way to express such a thought? In the unlikely scenario that milk drinkers ever *do* transform into piles of powder, how do we sound the alarm? We don't just need a default order of operations; we also need a way to override it. That's where parentheses enter the picture.

Execute the operations from most powerful to least powerful, *unless I say otherwise by using parentheses*.

So to add before multiplying, I can write $(2 + 3) \times 4$. Now the $2 + 3$ is a sealed-off unit, a capsule unto itself. We resolve the action inside the parentheses before we peer outside of them. The final result will be 20.

Or to multiply before exponentiating, I write $(2 \times 3)^2$. Now the 2×3 is a pair of numbers joined in parenthetical matrimony. No outside operation can come between them. The final result will be 36.

What if I want to add, then multiply, and *then* exponentiate? In that case I stack the parentheses, wrapping them like layers of packaging around the numbers, to make clear who belongs with whom. Instead of $1 + 4 \times 3^2$, I write $((1 + 4) \times 3)^2$.

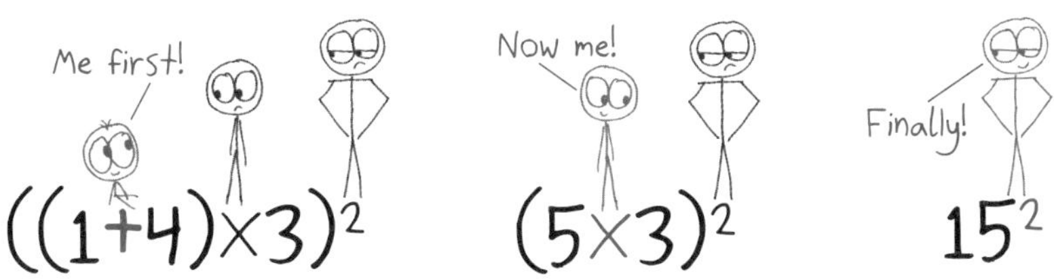

There are two kinds of rules in mathematics; we might call them *conventions* and *laws*. Conventions deal with how to use and interpret mathematical language: For example, the fact that 4½ is equal to $4 + \frac{1}{2}$. We could, if we desired, call a global meeting and all agree to change this convention, so that 4½ now means $4 \times \frac{1}{2}$. It would be like all English speakers agreeing that "friendship" now means "ice cream." Weird, but perfectly valid.

Meanwhile, laws deal with mathematical truth. Though expressed in language, they run deeper than language: For example, the fact that $a \times b$ is equal to $b \times a$. No global meeting could ever alter this truth. We can tinker with the language as much as we like, but whatever language we settle on, the same underlying principle will hold.

The order of operations is a funny case. It's a convention that's widely mistaken for a law.

Witness the peculiar genre of problem that goes viral on social media every few months. A typical example popped up as I was beginning work on this book, a question that mathematician Steven Strogatz, writing in the *New York Times*, called "artfully perverse, as if constructed to cause mischief."

That question: What is $8 \div 2(2 + 2)$?

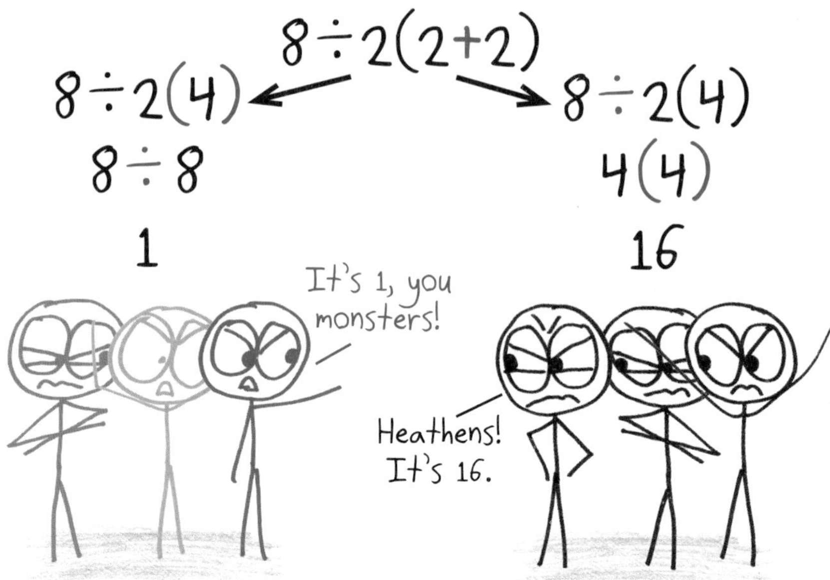

The trouble lies in what's missing, that absence of a symbol between the first 2 and the parenthetical. Such an absence signifies multiplication; that is, we are multiplying 2 by (2 + 2). But does this multiplication occur before the division or afterward? The division lies to the left, so following left-to-right logic, we ought to do 8 ÷ 2, arriving at 4, after which we should multiply by (2 + 2), to get 16.

But doesn't it seem strange to divorce the 2 and the (2 + 2) when there is not even a symbol separating them? What if the left-to-right rule applies only when multiplication is signified by an actual symbol, such as × or ∗? According to this logic, we ought to multiply 2 by (2 + 2) first, to get 8; then, carrying out the division, 8 divided by 8 gives 1.

So who is right: those who say 16 or those who say 1?

To be candid, I don't care. This isn't a question about laws; it's a question about conventions, and the conventions here are ambiguous. It's not like asking what will happen when a scientist drops a rock into a pond; it's like asking what a scientist meant by doing so.

Since mathematics is a language, when people use it poorly, there is only one solution: to ask them what they mean.

Calculation

One of my all-time favorite questions comes from teacher Claire Longmoor:

An orchestra of 120 players takes 40 minutes to play Beethoven's 9th Symphony. How long would it take for 60 players to play the symphony?

In recent years, Claire's query has drifted around the internet, drawing incredulity and rage wherever it lands (and not just because a proper performance of Beethoven's Ninth ought to take more than an hour). The question, as written, is nonsense. Dismissing a violist does not change a symphony's tempo. "That's not how this works," scoffed one person, in a tweet read by millions. "That's not how any of this works." Another chimed in with a comparison: "It takes nine months for a woman to have a baby; how long would it take for two women to have a baby?" It's easy to share their scorn. After all, what kind of teacher would ask such a bogus question?

A clever teacher, it turns out. Claire was making a simple, potent, and strangely elusive point: *sometimes, the correct calculation is not to calculate at all.*

Consider another problem, written decades before Claire's:

There are 125 sheep and 5 sheepdogs in a flock. How old is the shepherd?

You cannot answer the question with the information provided. Age is not denominated in the units of "sheep per dog." Yet in a classic study (and in subsequent replications), roughly three-quarters of elementary school students attempted some kind of calculation to solve this problem. Most divided the numbers to get 25. Others averaged to get 65, added to get 130, or even multiplied to get a spry shepherd with the biblical age of 625.

Many students knew this was silly. But they could not embrace Claire's wisdom. They could not resist the imperative to calculate.

Why does calculation matter to begin with? Well, because measurement takes us only so far. We cannot, for example, measure distances to stars. Instead, we measure what we can—say, the change in a star's angle of inclination in the sky from season to season—and then we calculate what we wish to know. Calculation transforms old numbers into new wisdom.

But only if it's the *right* calculation.

Here's a subtler example. Three customers at a restaurant run up a $25 bill. They each chip in $10 and receive $5 as change. Deciding to keep $1 each, they leave the remaining $2 as tip.

But hold on: Since each guest ultimately paid $9, that comes to $27. Then, the $2 tip brings us up to $29. Yet they began with $30.

What happened to the last dollar?

It feels like a real question. It seems to demand a real answer. Yet it's just as nonsensical as the ageless shepherd. The proposed "missing dollar" is only the phantom artifact of a meaningless operation. I added $27 (the total amount spent) and $2 (the tip), even though the $2 is already included in the $27. As Claire Longmoor would tell us, that's foolishness. If you're trying to account for the $30, then $25 went to the bill, $2 to the tip, and $3 back to the customers. The only mystery is why we carried out the totally irrelevant operation of 9 + 9 + 9 + 2.

We tell rude children: "If you don't have something nice to say, don't say anything at all." There's a mathematical equivalent: *If you don't have something sensible to calculate, don't calculate anything at all.* But as with holding back insults, it's easier said than done. Or easier said than *not* done. School mathematics is a flurry of activity: adding, subtracting, multiplying, dividing, not to mention finding distances, areas, volumes, factors…When you've spent years in a frenzy of motion, nothing is harder than stopping.

At the very beginning of this section, I pointed out that operations are not really verbs. The + in 2 + 3 works more like a conjunction (2 *and* 3) or a preposition (2 *with* 3). I called this a "seemingly minor technical point," and set it aside for later. Well, later is now. We have been reading + and – as verbs, telling us what to calculate. Now, we must read mathematics in a whole new way.

Not as *instructions* but as *structure*.

For example, here's the simplest operation I can think of: 1 + 1. We have been reading this as a command: "Add these numbers." But in the grammar of math, it is not a command. It is only a thing: two nouns linked to form a noun phrase. "One and one" is akin to "a dog and another dog." You may, if you wish, paraphrase this as "two dogs." But such a paraphrase is optional. In this language, 1 + 1 isn't a question whose answer is 2; it's a noun whose synonym is 2.

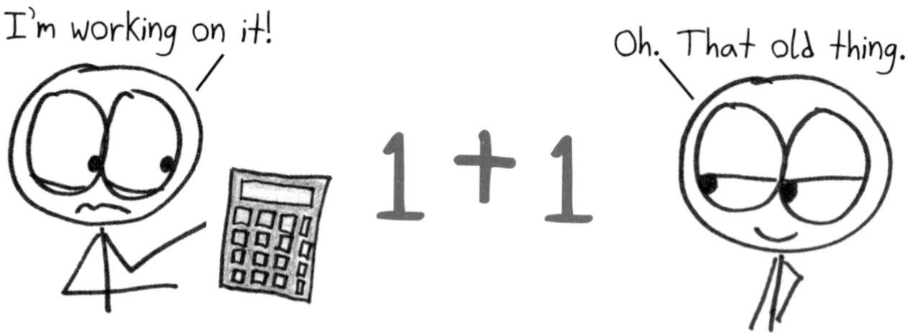

Here's a richer example: 3 × 7 × 11. This is not a command. It is a number. You can, if you like, replace it with the synonym 231, but something is lost in the process. Ask me to divide 231 jelly beans into equal-sized piles, and I'm at a loss. Ask me to divide 3 × 7 × 11 jelly beans into equal-sized piles, and I know just what to do: I can make three piles of 7 × 11, or seven piles of 3 × 11, or 11 piles of 3 × 7. To replace 3 × 7 × 11 with 231 erases all this information; only by refraining from calculation does the number's nature remain clear.

Sometimes, the correct calculation is not to calculate at all. Ignore the instructions and focus on the structure.

When I showed a draft of this book to my uncle Paul, this is where I lost him. "Not instructions but structure?" he said. "I don't get that. To me, math *is* instructions."

I know Paul's not alone. It's awkward to read 3×4 as a self-sufficient thing, a noun whose synonym happens to be 12. This way of reading mathematics (which will occupy the next section of the book) once had the fitting name of "cossism," after the Italian *cossa*, for "thing." In a word: *thingism*.

Still, I'll stick with the more familiar title: *algebra*.

÷10
10⁻¹ 10⁰ 10¹ 10² 10³
0.1 1 10 100 1,000
π ≈ 3.14
τ ≈ 6.28
y=mx+b
around
across
"857"
-(-(-(-1)))
Ah, simple.
125
+ 124
249
1/4 + 1/6 =
∀ε>0,
∃δ>0:
x–a|<δ→
f(x)–L|<ε
ax²
+++
THE FUZZY TRUTH
96°F
97°F
98.6°F
unreasonable rounding
CHOCODOLLAR
10X10
1/10X10X10
3.14
π
22/7
3.140 3.141 3.142 3.143
17 × whatever
1.414ish
PI DAY PARTY!
3/14
3.4104398
+ 2.5899496
6.0003894
5×3
multiplication
4/7
$10
farther
closer
2573
× 389
3 + -1.78
division by 1/2
17
34
division by 2
???
0 5 10 15 20 25 30 35 40 45 50
repeated addition
5⁵⁵
repeated exponentiation
5↑↑3
tetration
NOOOOOO
a÷b/c=
4⁷ᐟ² = ???
6
2
λ
x(x+x²)?
x²+x³!
8÷2(2+2)
8÷2(4)
8÷8
1
whisper
conversation
blender
nightclub
thunder
20 30 40 50 60 70 80 90 100 110 120
×1000
x²=3x+10
K=½mv²
If A<B and B<C
then A<C
x = −b ±√b² − 4ac / 2a
DONUTS
7 PIZZAS
10/6
44.9° North
93.2° West
ax² = bx + c
ax² + bx = c
ax² + c = bx
ax² = bx
ax² = c
bx = c
1/3 sleep deprivation,
2/3 game theory.
Guaranteed 100%
Almost certain 95% to 99.9%
Very likely 80% to 95%
Probable 60% to 80%
Maybe 40% to 60%
Possible 20% to 40%
Unlikely 5% to 20%
Almost impossible 0.1% to 5%
Impossible 0%
AROUND ≈ 3 × ACROSS
OOOOH!
17/34
noisy
"1000"
3
5
1 −1 −(−1) −(−(−1))
11
5↑↑3 WOW!
30×20 30×7 8×20 8×7
600 + 210 + 160 + 56 = 1026
5+3 5×3 5³
No! Silence your nonsense!
82,771
What is lim x→2 x²−4/x−2?
0 .1 .2 .3 .4 .5 .6 .7 .8 .9 1
What's 46 + 28?
let's see... calculating... calculating...
27

GRAMMAR

The Syntax of Algebra

I sympathize with grammar teachers. Their subject, like mine, gets a bad rap. Students tend to see it as a form of institutionalized nagging, designed to flatten young people's natural speech ("My friend and me...") until it's as lifeless as old people's ("My friend and I...").

As I understand it, this is almost precisely backward.

Consider a *pidgin*. A pidgin arises when speakers of different languages are thrown together and forced to communicate. They cobble together a working phrase book, an impromptu collection of stock expressions, drawn from their various languages. A pidgin is a practical necessity, but it is no one's first language and, indeed, not really a full language at all.

Then they have children. Children are exquisitely attuned to learn the surrounding language—but what if there *is* no surrounding language? In that case, the kids perform an effortless miracle: simply by speaking among themselves, they elaborate the pidgin into a system of full complexity. Just as the wooden Pinocchio turned into a living boy, the pidgin comes to life as a complete language, known as a *creole*.

The difference—what a pidgin lacks and a creole has—is *grammar*. The spontaneous speech of young people is not *opposed* to grammar; it *is* grammar.

In a word, grammar is structure. It's what allows scattered words to take linguistic life. Its rules are not like those of etiquette but those of chemistry, accounting for how tiny atoms (sounds, words, suffixes) combine to form the varied and inexhaustible matter of language.

So what is the grammar of math?

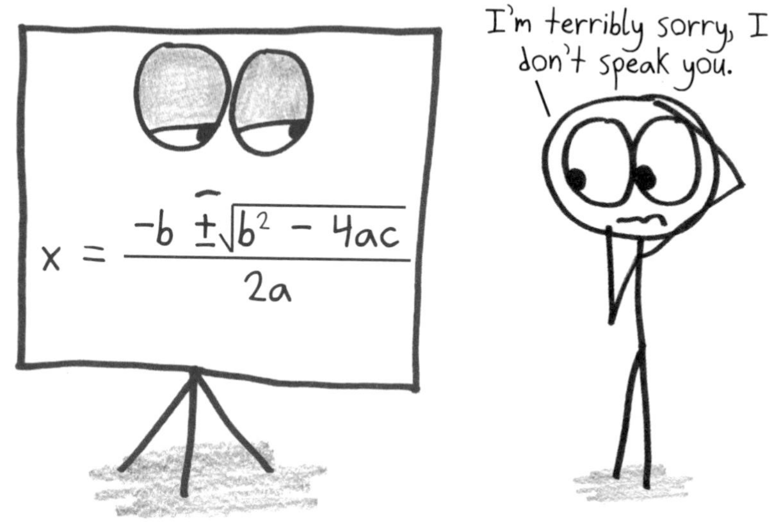

Thus far, this book has tackled *arithmetic*. Its nouns are numbers, its verbs are operations, and together, the two can express some practical, concrete thoughts, such as $9 = 2 + 7$.

But arithmetic is more of a pidgin than a language. Why do different operations sometimes give the same output? When can a tedious computation be streamlined? How does an inaccurate measurement affect a later calculation? Arithmetic raises these questions but cannot answer them (at least, not in the form we've learned it). The pidgin cannot comment on itself.

Until now. In this section of the book, the pidgin blossoms into a creole. The phrases of arithmetic birth the language of algebra.

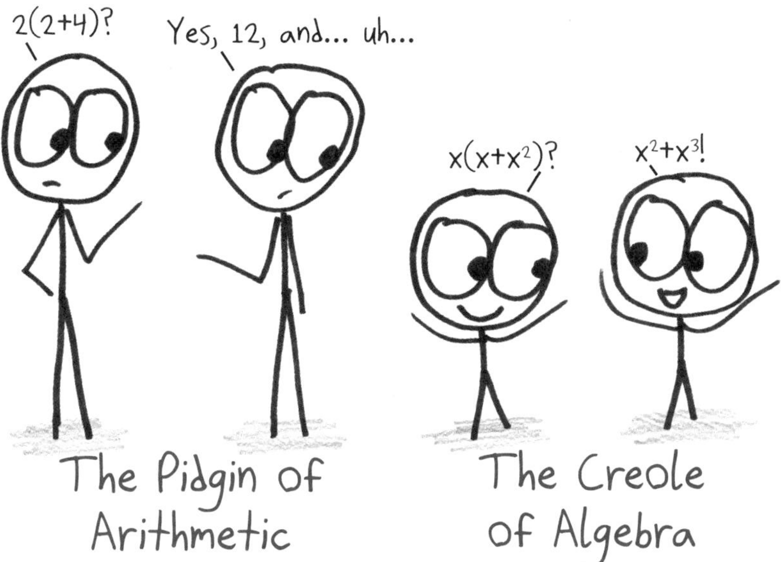

The word "algebra" comes from the Arabic *al-jabr*. The word refers to the reunion of broken parts, as in the healing of a fractured bone. It came to math as a metaphor: like a surgeon setting a broken bone, an algebraist reunites the fractured pieces of unknown quantities. We can also take the metaphor in a different direction: if arithmetic is a collection of bony fragments, then algebra fuses them into a cohesive whole, a new limb for the human mind.

Symbols

When I ask mathematicians for their favorite piece of notation, they tend to bristle. It's an understandable reaction: The question is a bit like asking a musician for their favorite note, or a chef for their favorite oven temperature. Not exactly honoring the finer points of the art. Yet when they deign to answer, mathematicians show a surprising consensus: these folks, they love their sigma notation.

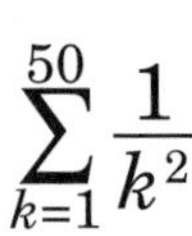

The word "symbol" comes from the ancient Greek *symbolon*: a piece of an animal's knucklebone, broken in half and then shared between two people, like one of those heart-shaped BEST FRIENDS FOREVER lockets. A symbolon linked you to a faraway companion; a symbol links you to a faraway concept. It is the token of a distant abstraction.

As such, the mathematical alphabet isn't quite like the English one. Every symbol names an idea. Instead of 26 letters expressing a few dozen sounds, we have thousands of symbols, their meanings in perpetual flux as we seek to capture an ever-expanding repertoire of ideas.

In math, our alphabet includes:

Subtly Different Vertical Lines

1 l i I || ‖

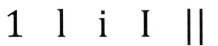

Subtly Different Circles

o O 0 θ Θ

Subtly Different Crisscrosses

x X × χ

Various Misshapen E's

Stuff We Crossed Out

Every Known Species of Bracket

Little Dot Clusters

  :

Off-Brand Equals Signs

Letters Wearing Fancy Hats

ÿ ŷ ỹ ȳ 

Words That Don't Mean What You Want Them to Mean

log tan sec sin

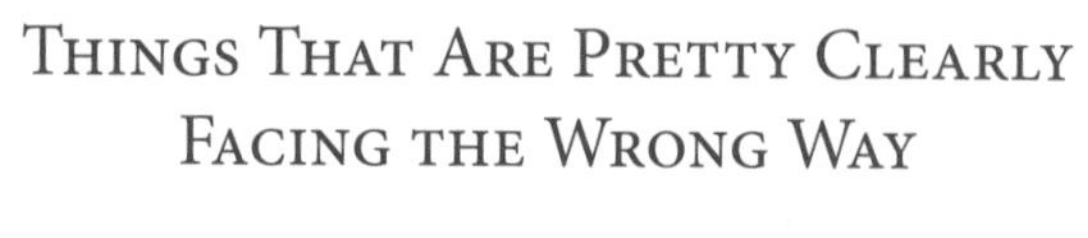

Things That Are Pretty Clearly Facing the Wrong Way

$\forall$ ∞ $\exists$ $\mp$

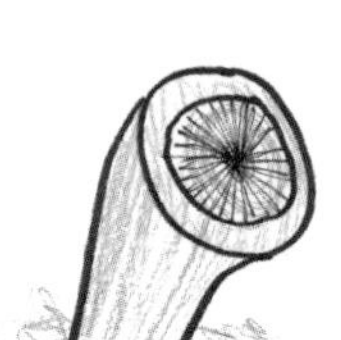

The Dune Logo

$\supset$ $\cup$ $\cap$ $\subset$

Too Many Ways to Say "Multiply"

$a \times b$ $a \cdot b$ $a * b$ ab

Ones That Just Don't Feel Trustworthy

$\propto$ Ξ $O(n)$

f^{-1} a_{k_3} $\sin^2 x$

To taste how math's alphabet works, let's unpack the little thicket of symbols we saw above:

$$\sum_{k=1}^{50} \frac{1}{k^2}$$

The imposing Σ (a capital Greek letter *sigma*) stands for "sum." Specifically, it adds together, in a single total, many copies of the thing to its right (in this case, $\frac{1}{k^2}$). The value of k begins with the number below the sigma (in this case, $k = 1$) then clicks upward one step at a time (2, 3, 4) until it arrives at the number atop the sigma (in this case, 50).

Here's the same quantity, unspooled into an ordinary sum:

$$\frac{1}{1^2}+\frac{1}{2^2}+\frac{1}{3^2}+\frac{1}{4^2}+\frac{1}{5^2}+\frac{1}{6^2}+\frac{1}{7^2}+\frac{1}{8^2}+\frac{1}{9^2}+\frac{1}{10^2}+$$

$$\frac{1}{11^2}+\frac{1}{12^2}+\frac{1}{13^2}+\frac{1}{14^2}+\frac{1}{15^2}+\frac{1}{16^2}+\frac{1}{17^2}+\frac{1}{18^2}+\frac{1}{19^2}+\frac{1}{20^2}+$$

$$\frac{1}{21^2}+\frac{1}{22^2}+\frac{1}{23^2}+\frac{1}{24^2}+\frac{1}{25^2}+\frac{1}{26^2}+\frac{1}{27^2}+\frac{1}{28^2}+\frac{1}{29^2}+\frac{1}{30^2}+$$

$$\frac{1}{31^2}+\frac{1}{32^2}+\frac{1}{33^2}+\frac{1}{34^2}+\frac{1}{35^2}+\frac{1}{36^2}+\frac{1}{37^2}+\frac{1}{38^2}+\frac{1}{39^2}+\frac{1}{40^2}+$$

$$\frac{1}{41^2}+\frac{1}{42^2}+\frac{1}{43^2}+\frac{1}{44^2}+\frac{1}{45^2}+\frac{1}{46^2}+\frac{1}{47^2}+\frac{1}{48^2}+\frac{1}{49^2}+\frac{1}{50^2}$$

Add it all up, and it equals about 1.625. Or, if you prefer your answers unrounded, it equals precisely

$$\frac{3,121,579,929,551,692,678,469,635,660,835,626,209,661,709}{1,920,815,367,859,463,099,600,511,526,151,929,560,192,000}.$$

Impressive, no?

As we'll explore in the next chapter, the variable k is a triumph of concision. Whereas a number like 4 or 7 has a single definite meaning, the slippery number k does not. In the above phrase, it takes on 50 meanings (1, 2, 3, and so on) in the span of a single instant. The compression is marvelous.

Meanwhile, the Σ achieves a different kind of compression. Whereas k expresses 50 simple ideas, Σ expresses one complex idea: *add together these items in a single sum.* Such concision is common in math, where a few funky marks—say, $SL_n(k)$ or $O(n^3)$—can encapsulate an idea so intricate it requires several years of undergraduate education to grasp. It's as if a single musical note expressed the whole of Beethoven's Ninth, or a single letter the complete text of *Sense and Sensibility*.

Anyway, when the symbols Σ and k team up, the result is something special. Fifty fractions in just nine marks. It's a kind of mathematical clown car, packing dozens of clowns into a space scarcely larger than the words "dozens of clowns." Even more striking, we could (by changing the number atop the sigma) just as easily make it 50,000 clowns—or 50 billion.

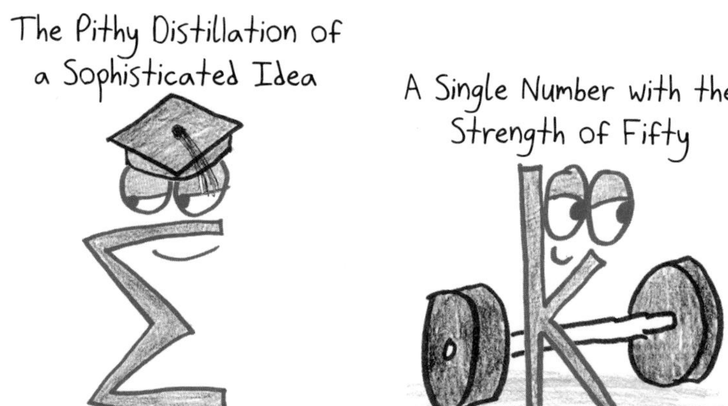

The Pithy Distillation of a Sophisticated Idea
A Single Number with the Strength of Fifty

A colleague of mine once asked in frustration why her students refused to read the math textbook. Her background was in biology, a subject where the book—dense and difficult though it may be—is an irreplaceable source of learning. Now she was teaching algebra and was losing patience with her students' incapacity to glean anything—anything at all—from the text. "Why do they need it spoon-fed to them?" she asked me.

I had to confess that I'm not much good at reading math myself. As a psychology/mathematics double major, reading a 10-page psych paper would take me 20 to 30 minutes, whereas a 10-page math paper might require a couple of days. Or, for that matter, decades.

The obstacle is not the density of the readers; it's the density of the symbols.

To learn English, you begin with the ABC's. In math, that's just not possible; you learn the ABC's as you go. The result is a language that's painstakingly slow—and uncommonly beautiful. On inspection, an unassuming little k or Σ turns out to be a precious family jewel, concentrating worlds of value into a tiny space.

That's another reason mathematicians might struggle to pick a favorite notation: when every symbol is laden with significance, it's hard not to count them all as favorites.

Variables

Let me guess: the moment that math lost you was "when the letters got involved."

Perhaps this is not specifically true of you. Perhaps you are, for example, Pulitzer Prize–winning journalist Natalie Wolchover of *Quanta Magazine*, in which case, I love your work, Natalie! Still, Pulitzer or not, I'm sure you've heard the timeless lament "I was okay with numbers, but I couldn't handle the letters." I myself have heard too many variations on this theme to count them all—unless I'm allowed to count using letters, in which case, I've heard it n times.

So what's the deal with n? Or for that matter, with x, y, m, and q, not to mention their Greek cousins θ and ε and μ and the like?

What they are, in a word, is *variables*.

What they do, in a word, is *vary*, from one value to another.

But my advice is this: *think of them as mathematical pronouns.*

English pronouns come up all over the place, but I'm referring to a particular usage: pronouns as personal placeholders. For example, instead of saying "Ben is a nuisance," you may say, "*He* is a nuisance." The pronoun "he" substitutes for the noun "Ben." Pronouns like "he," "she," and "they" let us refer to nuisances without mentioning their names—indeed, without even *knowing* their names.

Likewise in math, variables let us refer to unnamed or unknown numbers. $3 + x$ is shorthand for "three more than this other number." You might say: "three more than *her*."

3 + −1.78
3 +
Don't mind me. I'm just a stand-in.

A variable, like a pronoun, often refers to a specific entity. If I ask whether a singing mermaid has passed by here, you may reply, "Yes, she swam that way." The *she* is someone specific, whose name you don't happen to know (though Ariel is a fair guess). Likewise, if we say $3 + x = 5$, then x refers to one number in particular whose name we haven't yet uttered (though you may have a hunch).

Still, variables have a greater power: They let us generalize. A single variable may refer to many numbers all at once.

In English, we tend to generalize via plurals. "Cows eat grass." "Nations need wise leaders." "Follow-up albums are rarely as good as debuts." But in mathematics, there are no plurals. We must use the singular instead. "A cow eats grass." "A nation needs a wise leader." "A follow-up album is rarely as good as its predecessor."

Singular pronouns can serve a similar role. For example, to forbid dangerous behavior among the toddlers I know, I can write up a list of specific prohibitions: *Casey must never run with scissors. Ronja must never run with scissors. Rayhaan must never run with scissors*...But it's easier and more comprehensive to issue a blanket rule: *one must never run with scissors.* The word "one" is a pronoun, a generic placeholder that stands in simultaneously for Casey, Ronja, Rayhaan, Rachel, Hanan, and all the adorable others.

So it is with mathematical rules. I may say "One pizza feeds three people," "Two pizzas feed six people," "Three pizzas feed nine people," "1 billion pizzas feed 3 billion people," and so on. But the list never ends. Easier to say: "Any number of pizzas will feed three times that number of people." Or better yet: p pizzas will feed $3p$ people.

This variable p is a marvel of concision, compressing an infinite list of statements into just one.

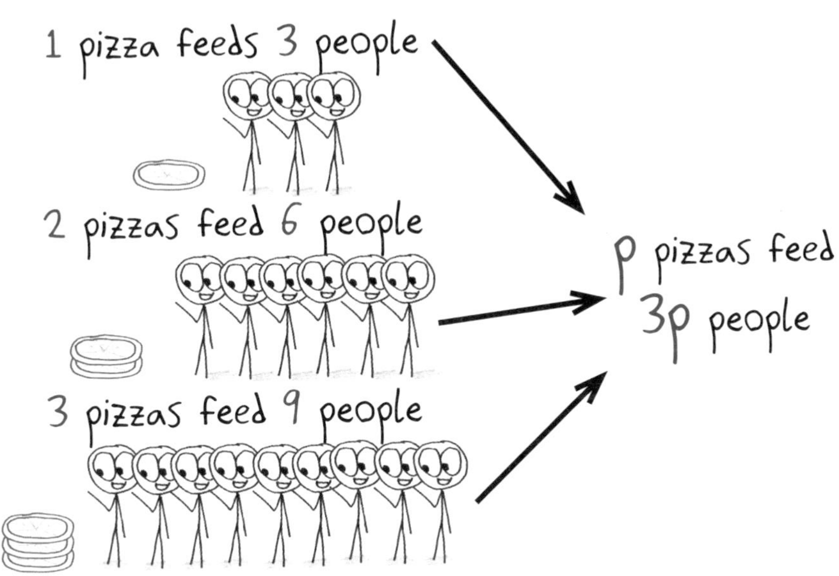

With English pronouns, there is a risk: unclear antecedents. For example, in the sentence "He told him his password," whose password are we talking about? Both the password teller and the password hearer share the same pronouns ("he"/"him"), so "his" could refer to either of them. The antecedent—the person to whom the pronoun refers—is ambiguous.

One can imagine the same confusion arising in math. "One number equals another number plus its square." Whose square: the first number's or the second number's? We avoid this trouble by assigning each number its own unique variable. It's a language so flexible that each person has their own pronoun. "One number" becomes x, "another number" becomes y, and the full equation, depending on whose square we mean, is either $x = y + x^2$ or $x = y + y^2$.

This multiplicity of pronouns raises the question of which one to pick for any given number. The answer (unlike in English) is that it's up to the speaker. You can assign a number whatever pronoun you want.

For example, I said earlier that p pizzas feed $3p$ people. Why p? No deep reason; it's just that "pizza" starts with p. The letter is only a placeholder; it carries no meaning in itself. I could equivalently have said that x pizzas feed $3x$ people, or that β pizzas feed 3β people, or that 🔥 pizzas feed 3🔥 people. The sentences all communicate the same thought.

I once taught a middle school class suffering from a plague of bad penmanship. The students couldn't read their own handwriting: They mistook b's for 6's, g's for 9's, and t's for +'s. A few of them, in desperation, would replace all variables with x and y, these being the only symbols they could reliably discern. Such a substitution is valid, if somewhat antisocial. It's as if someone said, "Here's a story about two fictional people, Alice and Bob," to which you reply, "No, I'm calling them Xochitl and Yusuf." It's clear what you mean, but it adds some friction to the communication.

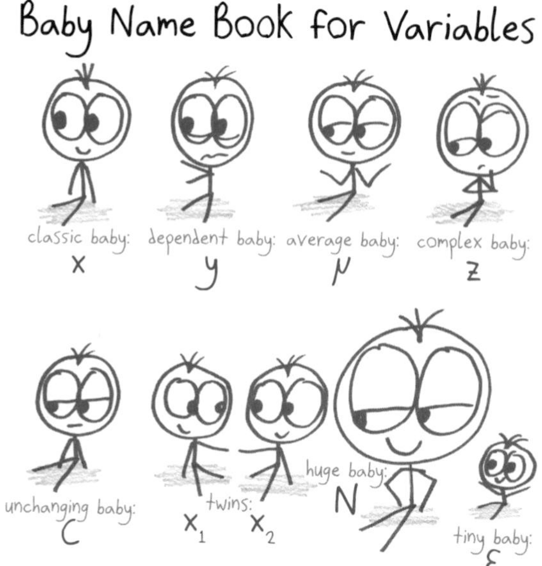

Variables are tools for communication. That's why mathematicians have developed communal habits for choosing variable names. These shared conventions help us to build shared knowledge.

No one is twisting your arm to comply. You may, if you like, name a large number ε and a tiny one N. But it'd be like naming a flooring company Jillian, and a child Flooring Solutions Inc. Perfectly legal, just wildly confusing.

I understand why folks find the letters unsettling. But mathematics without variables would be like English without pronouns. Just imagine: I could address this book to Natalie Wolchover, or to my former students Shizhou and Kisa and Lokesh, or to singer-songwriter Josh Ritter, or to any number of people…but I could never have written it for *you*. (Unless, of course, you happen to be Josh Ritter. In which case, hey, man, I love your music!)

Expressions

As a student, I couldn't make heads or tails of Ernest Hemingway. The terseness, the mud, the splashes of Spanish—it all baffled me (and I actually speak Spanish). When I mentioned my struggles to my friend Mike, he nodded. "Someone told me that before you can read Hemingway," he said, "you need to have done three things: gotten blackout drunk, been in a fistfight, and fallen in love."

I'm not 100% sure about the specifics, but Mike's larger point stands. To read literature, you need experience. Before reflecting on life, you have to do some living.

So what experience do you need, what kind of living is required, before you can read $4n + 2$?

In English, an "expression" is a familiar cliché, something like "Every cloud has a silver lining" or "You win some, you lose some." Those two happen to be full sentences, but other expressions are only phrases, playing the grammatical role of nouns: for example, "a perfect storm," "a blessing in disguise," or "the last straw." These latter examples point the way to *algebraic expressions*.

In math, an expression is a description of one number in terms of another. It's like saying "someone's favorite singer" or "someone's estranged chiropractor." For example, $x + 3$ is a number that happens to be three steps larger than another. Similarly, $5x$ is a number that's five times larger than another. As with Hemingway, it's easier to pronounce these brief expressions than to understand them. It's easy to read them yet hard to *read* them.

I first taught algebra to 11-year-olds in England's West Midlands. One of our first tasks was to explore sequences of numbers, like this one:

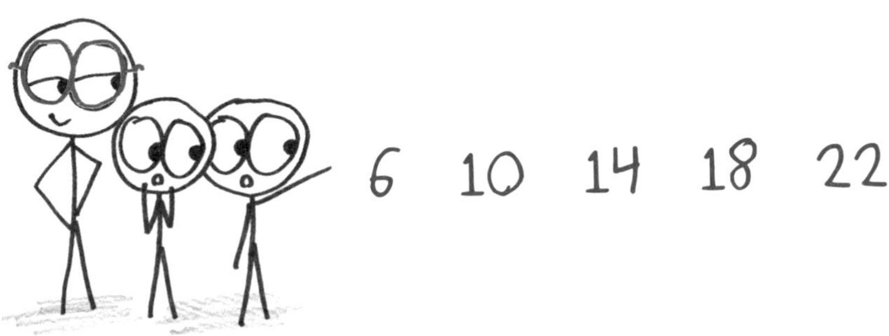

The goal was to develop a method for calculating any term in this sequence—the 7th, the 7000th, even the 7,000,000th—and then distill this method down to a pithy expression. To create such an expression was to solidify a process into an object, a *how* into a *what*.

I readied myself for a long and arduous lesson. But to my surprise, several of the boys cried out, "I know how to do this!" and promptly wrote:

How had they reached this expression? "You take the second number minus the first," was one boy's breathless explanation. "You put that next to the n. Then you subtract that from the first number, and you add on that result." The others nodded in fierce agreement.

But hold on: How exactly did this dense knot of algebra relate to the original string of numbers? What did n even mean? At these inquiries, they just blinked. To them, the task was complete, and my questions were as impertinent as "What is time?" or "Is anyone truly good?"

As I saw it, their grasp on algebra was no better than mine on Hemingway. So over the next few weeks, I developed a routine for explaining how to arrive at $4n + 2$.

First, we would look at the sequence together and enumerate the terms:

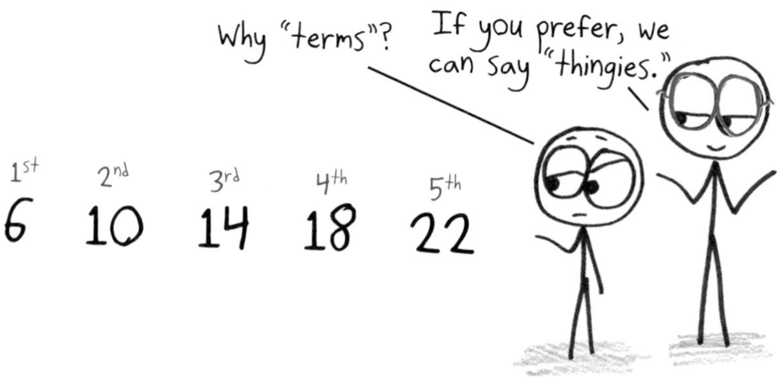

I'd ask what was happening at each step. Most students recognized a repeated process of adding 4.

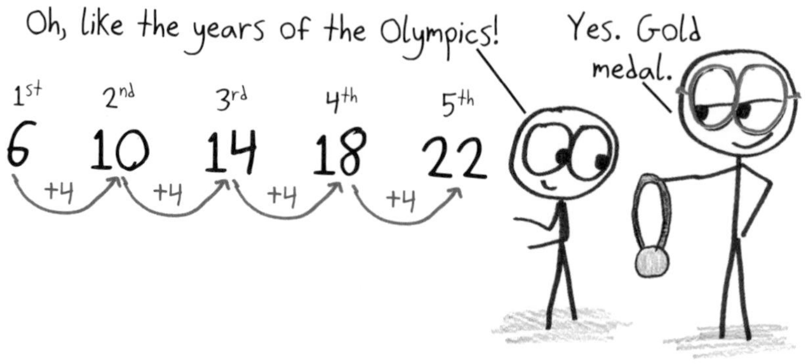

That allowed us to rewrite the sequence like this:

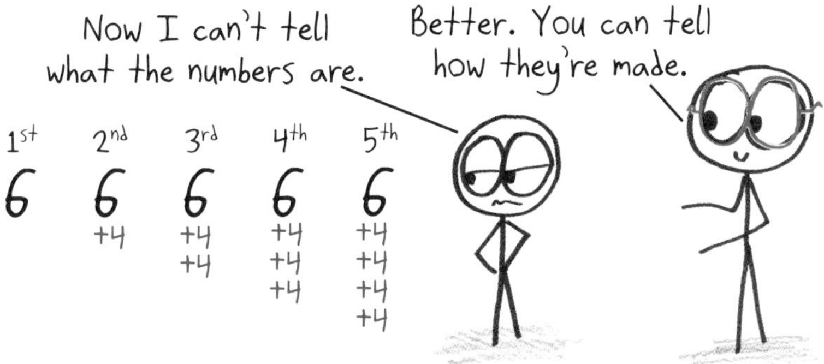

That's all well and good—but wouldn't it be nicer (I asked leadingly) if the first term had one 4, the second two, the third three, and so on?

My pupils, perhaps just to be polite, would agree, and so I'd produce that extra 4 by extracting it from the 6:

All of this was a mere prelude to the moments that followed. I'd begin a kind of rhythmic chant, a tuneless song of numbers.

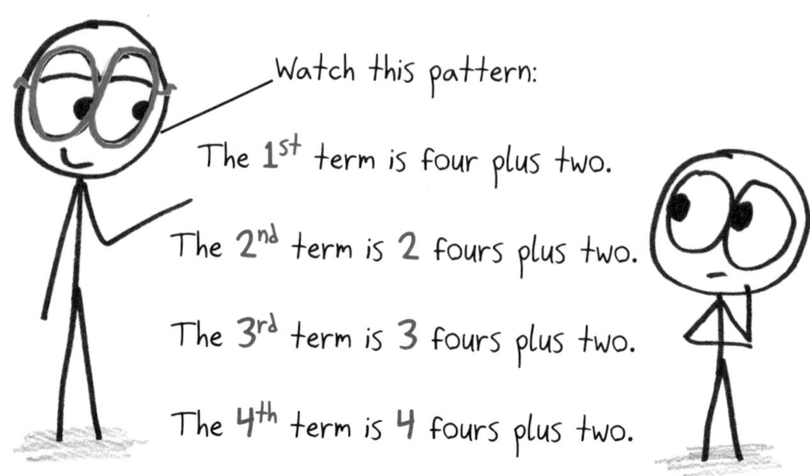

Watch this pattern:
The 1st term is four plus two.
The 2nd term is 2 fours plus two.
The 3rd term is 3 fours plus two.
The 4th term is 4 fours plus two.

By now, some students could wrap their mental fingers around the pattern. But in most cases, we needed more examples, more life experience. It became a game: I'd name a term, and they'd state its value.

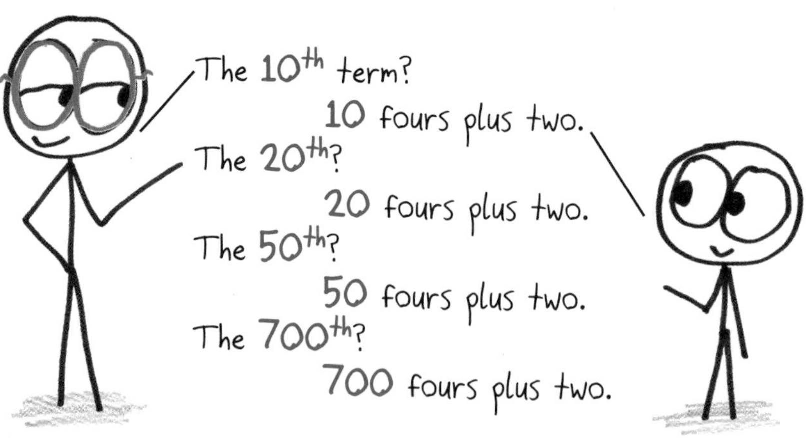

Never mind the precise results (which happen to be 42, 82, 202, and 2802). The pattern was taking shape. The outer layers of the sequence were falling away like clods of dirt. Beneath them we could glimpse a shiny metallic core. Onward...

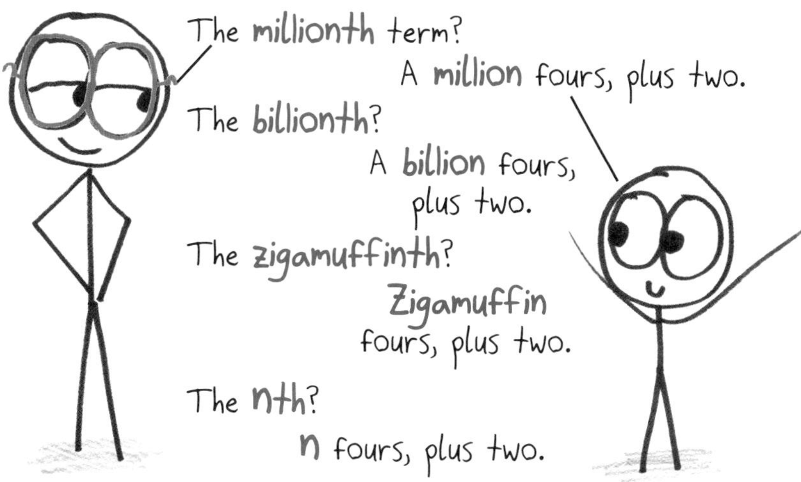

And there it was, born from a swirl of rock and dust like a newborn star. The *n*th term: *n* 4's plus 2. Or, if you will, $4n + 2$. Encoded in that little quartet of symbols was the recipe for calculating any term: the 7th, the 7000th, even the 7,000,000th. Our expression knew the terms, one and all, for the whole endless duration of the sequence. This nugget of algebra told an infinite story.

This is the algebraic leap: We move past operating on specific numbers. Now we *describe* those operations using *generic* numbers. Such mathematics, writes Karen Olsson, is "an early step back from numbers themselves, a shift of emphasis toward the dynamics of it all, from the stolid nouns to the freewheeling verbs that connected them."

An infinitely long sequence collapses into a single expression, radiating terms the way a star radiates photons.

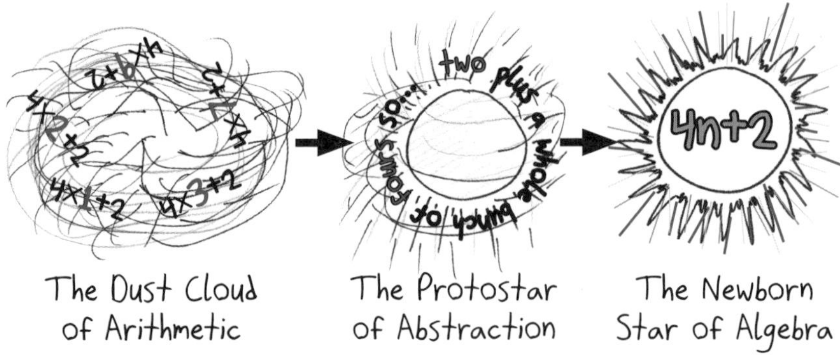

I never figured out how to conduct this as a whole-class lesson. The pace of learning is too idiosyncratic, too unpredictable. Instead, I ran this routine again and again, once for each student. Sometimes it took a dozen examples, sometimes several dozen, and sometimes just three or four. There is no universal threshold of experience at which math clicks, no moment when the process ("Multiply by 4, then add 2") suddenly crystallizes into a noun ($4n + 2$). We learners are maddeningly and beautifully individual. Perhaps you enjoyed Hemingway by age 15, and perhaps when I'm 75 I'll get in a drunken brawl and, when I come to, I'll finally appreciate *A Farewell to Arms*.

Equations

When jogging from Saint Paul to Minneapolis, I sometimes pass demonstrators: half a dozen sixtysomethings making their stand on a noisy bridge above the Mississippi and holding in their mittened hands a set of protest signs declaring their commitment to . . . Well, I'm not sure what. The writing is legible, but the meaning is vague. All I can really say with confidence is that they seem suspicious of war.

Nevertheless, I give a salute of solidarity. I, too, am suspicious of war, and I, too, admire the abstract symbol adorning the one sign whose specifics I can remember.

In English, there are three basic kinds of sentences. *Declaratives* are statements ("The sky is blue"); *interrogatives* are questions ("Is the sky blue?"); and *imperatives* are commands ("Go paint the sky blue"). As for math, the typical student tends to see it as a language of interrogatives (*What's the area?*) and imperatives (*Solve this equation*), with scarcely a declarative sentence in sight.

But the opposite is true. Every equation is a declarative sentence, a statement that two things are the same. The equals sign is not an exclamation point (*Calculate!*) or a question mark (*What does it equal?*), but a present-tense verb.

The symbol = is shorthand for "is equal to."

The grammar of mathematics is thus a bit repetitive. Translated into English, a page of equations would be a monotonous litany: "*A* is equal to *B*. Meanwhile, *C* is equal to *D*. Moreover, *E* is equal to *F*..." But in the native tongue of mathematics, this repetition isn't bad writing. It's crisp and efficient. With the sentences all sharing the same form, our minds are free to focus on their wonderfully varied content.

So what *is* that content? What do equations say?

To be frank, some equations (like some declarative sentences) are kind of banal. They're the mathematical equivalent of "I already ate lunch" or "My brother's name is Frederick."

Other equations are brazen lies. Take $x = x - 1$. Since a number cannot be one less than itself, this is categorically untrue.

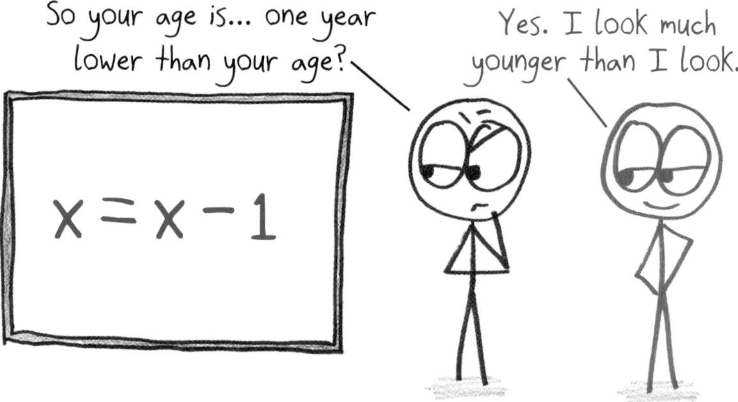

Other equations are *conditional*: true for certain values of the variable, but not for others. They allude to specific yet unnamed numbers, like bits of algebraic gossip. The equation $7 + x = 11$ tells us that seven and *some number* add up to 11. It's like saying "I saw *someone* with their arm around your cousin" or "It seems that *someone* ate the last cinnamon roll."

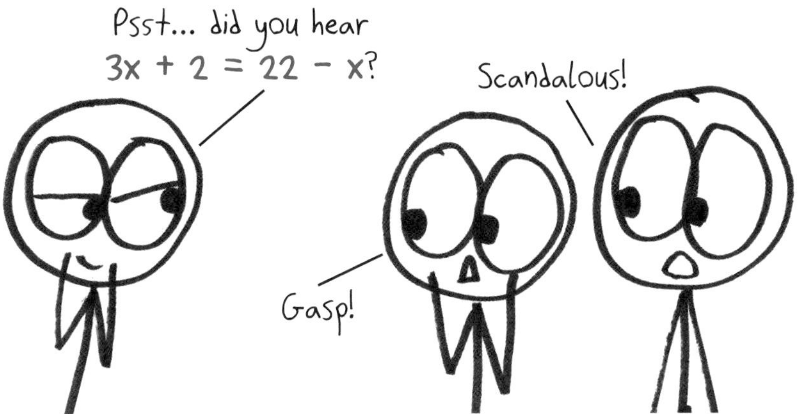

Finally, some equations are true for all values. We call such statements *identities*, and they are essentially platitudes: sweeping generalities, numerical clichés. In a sense, this makes them uninformative; but then again, as the saying goes, "Never underestimate the power of a platitude."

Now, having catalogued the varieties of equations, we could end the lesson here and jog home—if not for the matter of the "Ta-da!" problem.

Consider this incomplete utterance: $7 + 2 = \_ + 3$

What fills the blank? Well, reading the equation as a sentence, it translates to "7 + 2 is the same as __ + 3." That's easy enough; since the left side equals 9, the missing number is 6.

But students often write something else:

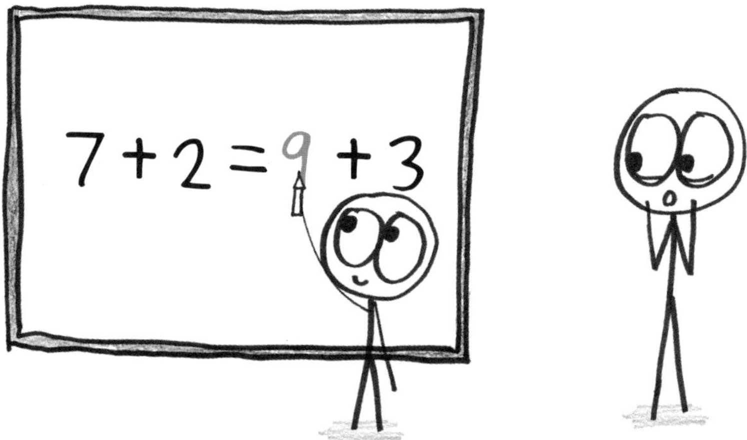

The sentence now says $9 = 12$. Not the truest claim I've heard. Then again, they're not reading the sentence as a claim. To these students, the equals sign is not a verb, but a kind of drumroll introducing the answer, a mathematical "Ta-da!"

This is the hallmark of pre-algebraic thought: seeing an equation not as a statement, but as a kind of action. The phrase $7 + 2 =$ leads automatically to the answer 9, as if we were punching numbers into a calculator in your skull.

It's not just kids who do this. Many people carry the "Ta-da!" view into adulthood, reading algebra not as a series of statements, but as a series of inscrutable instructions. If $8 - 5 = \_$ tells me to subtract, then what is $2x = x^2$ telling me to do? Hyperventilate? Scream? Major in the humanities?

In reality, $2x = x^2$ isn't telling you to *do* anything. As a conditional equation, it's sharing a juicy rumor: "Some number's double is also its square." The human equivalent would be "Somebody's dentist is also their lover" or "Somebody's veterinarian is also their mortal foe."

Of course, it's natural to be curious *who* we're talking about. Indeed, much of algebra consists of techniques for teasing out information from conditional equations so as to identify the mysterious number. The process is much like solving a whodunit, and fittingly, it is known as *solving* the equation.

But remember, the equation itself never asks you to lift a finger. It's only a statement of fact.

The protesters on the bridge seem to grasp this. Halfway between the Twin Cities, they hold a banner of twin concepts. Foolish joggers plod by, misconstruing their sign as a directive, a command, a call to action. But the protesters on the bridge know: "We are all equal" is only a plain statement of truth.

Inequalities

I was 15 when a friend told me the "half-plus-seven rule": the youngest person you should consider dating is half your age plus seven years. Right away, I began mentally plugging in ages, calculating the outcome, and weighing each result against my own intuition—an infallible authority, unmuddied by the experience of actual dating.

For example, a 20-year-old can date a 17-year-old, but not a 16-year-old. Seemed about right.

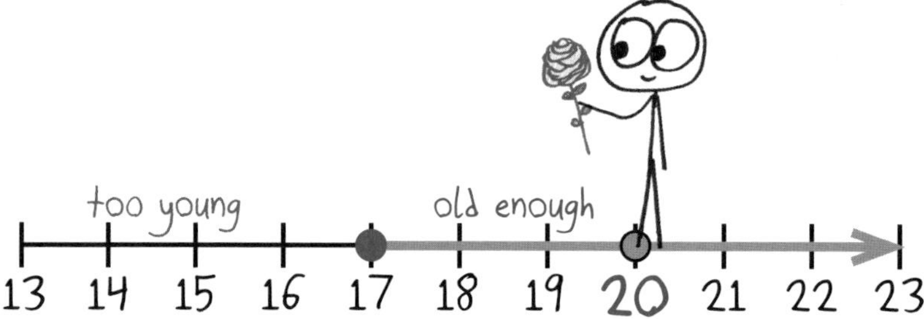

Meanwhile, a 26-year-old can date a 20-year-old, but not a 19-year-old. Fair enough.

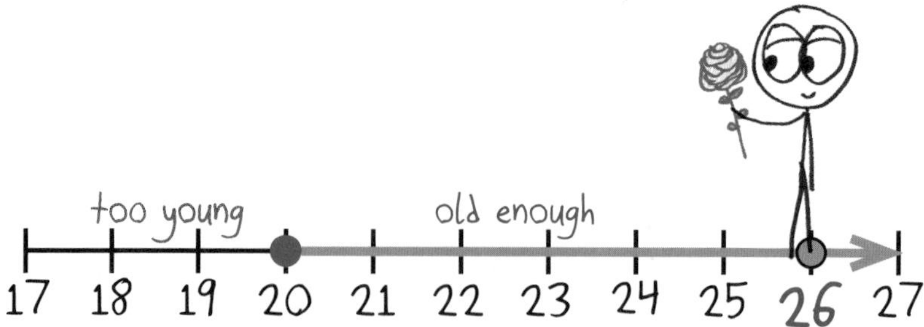

Also, a 40-year-old can date a 27-year-old, but not a 26-year-old. No argument with that.

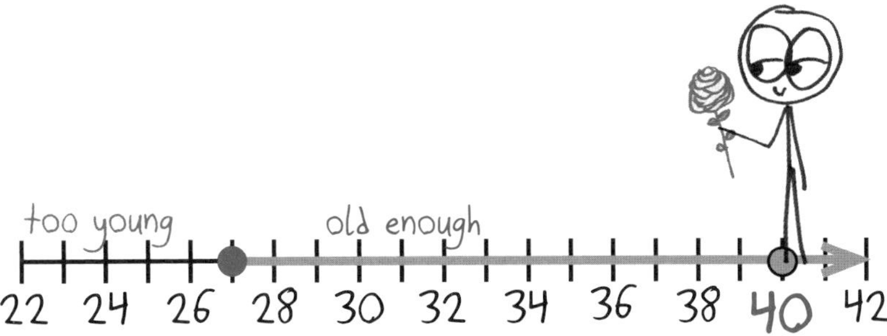

Lastly, if you're younger than 14, then you can date…no one at all. The rule would require a partner older than yourself, but for such a partner, you'd be impermissibly young. Perhaps that's why I liked the half-plus-seven rule: By its logic, dating cannot start until age 14. I wasn't a late bloomer; I was only following protocol.

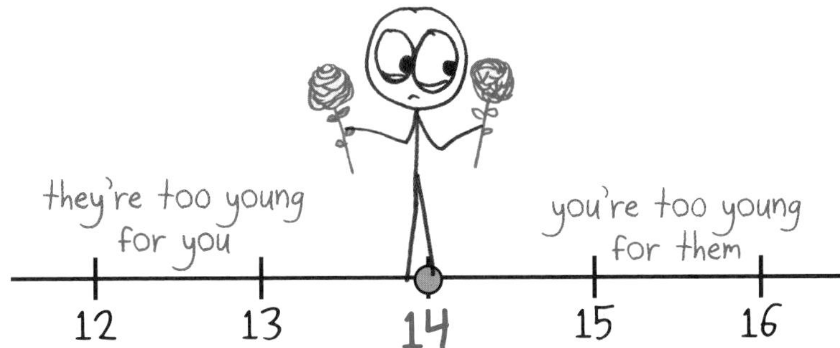

The half-plus-seven rule is an *inequality*. Whereas an equation declares two things equal, an inequality declares one thing larger (or smaller) than another. Anytime we say "more than," "less than," "at least," or "at most," we're speaking in inequalities. They appear all over the place in statistics, engineering, and beyond. Inequalities are the language of constraints, limitations, and tolerances.

Whereas equations all use the same verb, inequalities deploy four. First, $x > 4$ means x is greater than 4. It may be just a tiny bit greater (even 4.0001), but it cannot be 4 itself.

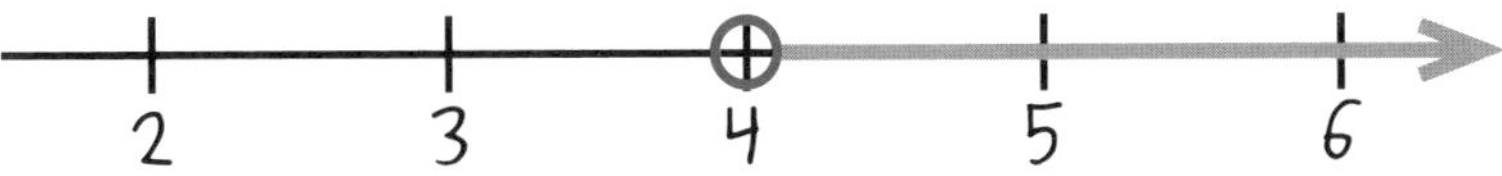

Next, $x \geq 4$ means x is *at least* 4. Four itself is allowed. We think of that underline as the bottom half of an =, and we often pronounce the symbol as "greater than or equal to."

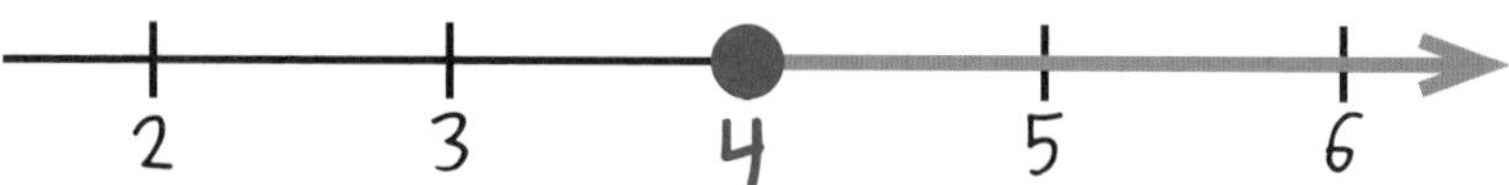

Third, $x < 4$ means x is less than 4. This includes 3.9999, but not 4 itself.

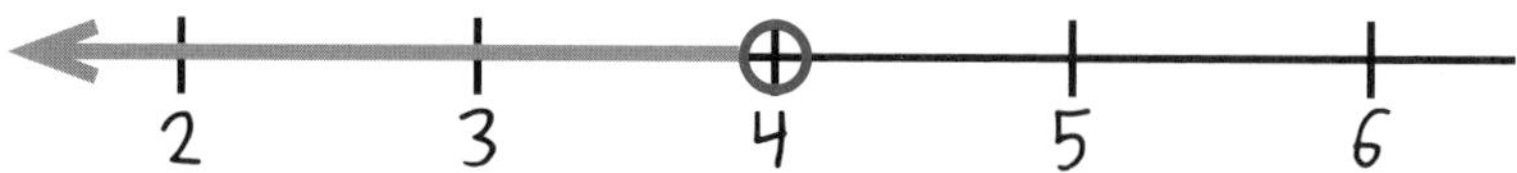

Finally, $x \leq 4$ means x is *at most* 4. It is often read as "x is less than or equal to 4."

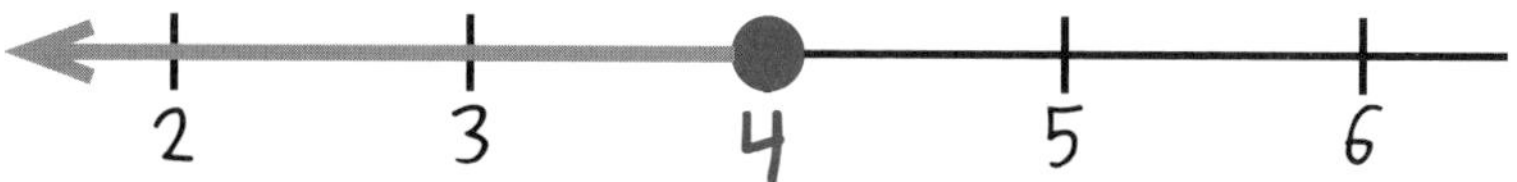

Inequalities tend to be more permissive than equations, in the sense that they allow for lots of possible values. Whereas $x = 4$ spells out a specific number, $x > 4$ allows for infinite possibilities, from 4.3 (and below) to 4.3 trillion (and above). Such freedom can be exhausting. Equations are like a partner who demands a specific restaurant, whereas inequalities smile and say, "Oh, I'm flexible, where would *you* like to eat?" More options for you—and more work.

Then again, such flexibility is often just what we need. A half-plus-seven *equation* would spell out a precise age gap for couples: 60-year-olds must date 37-year-olds and no one else. The half-plus-seven *inequality*, by contrast, allows for a range: a 60-year-old can date a 37-year-old, a 46-year-old, a 51-year-old, and more.

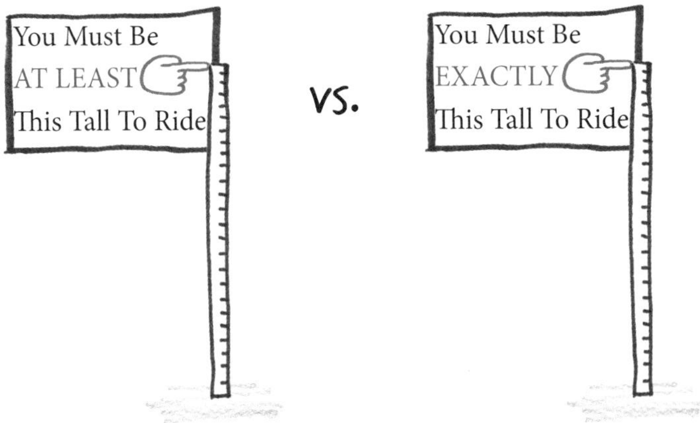

"People have the mistaken impression that mathematics is just equations," the physicist Stephen Hawking once said. "In fact, equations are just the boring part of mathematics." I suspect he was alluding to the geometric ideas housed inside the equations, but a certain breed of mathematician might say: "Yes, exactly! The interesting part of math is the inequalities."

Still, equations dominate popular visions of mathematics. No one writes books titled *The Happiness Inequality*. The word "inequality" evokes left-wing protests, not engineers calculating bridge tolerances. Yet the language of mathematics would be a hopeless mess without inequalities. "Nothing beats a good inequality in trying to understand a problem," mathematician Cédric Villani once wrote. "An inequality expresses the domination of one term . . . by another, of one force by another, of one entity by another." Even if equations fill our discourse, it's inequalities that fill our lives.

For example, a speed limit means that you can drive *at most* 65 miles per hour. No need to drive 65 on the dot.

Similarly, when my daughter asks for "five more minutes" at the playground, she's not demanding *exactly* five. Six or seven (or a hundred) would suit her fine.

Here's a favorite example of mine: If you're running errands on the way home from work, it's always shorter to go straight from A to B than to make a stop along the way at C. This innocuous fact, known as the *triangle inequality*, is fundamental to how mathematicians conceptualize distance.

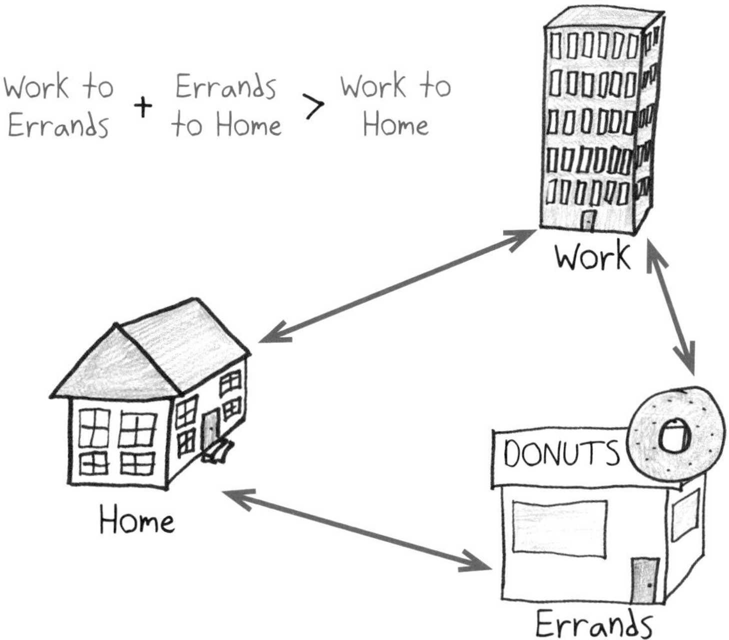

Inequalities don't have much hold in the popular imagination. They tend to languish in the shadow of those beautiful symmetric-faced siblings of theirs: equations. I often find myself speaking up for inequalities, as if the two kinds of mathematical sentences were in competition. But of course, that's silly. They work together; the key to solving an equation is often an inequality, and vice versa. I suppose this is the lesson of my own experience with the half-plus-seven rule: my wife, Taryn, and I are the same exact age, to within a week. In algebraic terms, you could say that the inequality $y > \frac{x}{2} + 7$ is satisfied by the equation $y = x$.

At least, for $x > 14$.

Graphs

When a student is stuck on a math problem, we math teachers often give the same advice: "Draw a picture." There's just one problem with this maxim: it's not always clear what picture to draw. My friend Michael Pershan once gave this advice to a third grader adding $\frac{1}{4} + \frac{2}{3}$. He hoped the kid would sketch a pair of shapes and then carve them into pieces. But when he circled back five minutes later, his student was not dividing a circle into quarters: he was adding signage and shading to a vivid illustration of a bakery. Draw a picture, indeed.

Such confusion is not limited to the youth. Adult textbook authors have been known to include gratuitous fighter jets and non sequitur cheetahs, as if the secret to math is picturing something, anything. But the hard part of math isn't remembering what cats look like. It's making sense of abstract ideas.

The challenge of algebra is to visualize the invisible.

Algebra is the study of relationships. Not in the romantic sense, but the sense of "two numbers that relate." The temperature in Fahrenheit relates to the temperature in Celsius. The diameter of a pizza relates to the number of people it will feed. The time it takes to go a mile relates to the speed at which you travel.

If we want to spell out these relationships in full generality and precision, we use *equations*. Equations encode a tremendous amount of information, but they can be hard to decode.

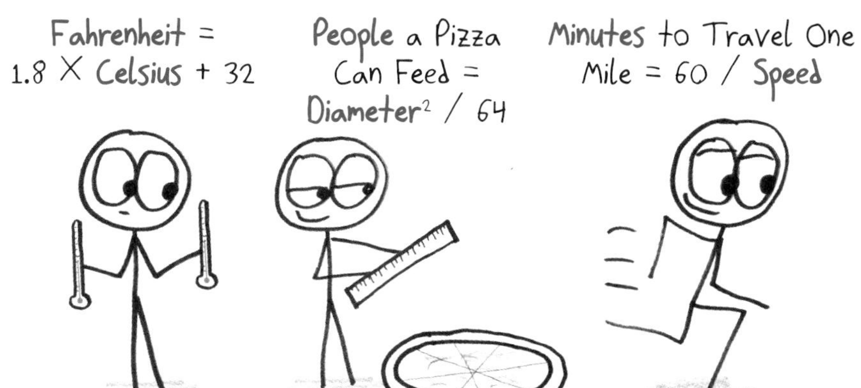
Fahrenheit =
1.8 × Celsius + 32
People a Pizza
Can Feed =
Diameter² / 64
Minutes to Travel One
Mile = 60 / Speed

Meanwhile, if we want to describe a relationship through a handful of clear and concrete examples, we use *tables*. Tables are easier than equations to take in, but they're incomplete—like a few thumbnail images drawn from hours of video.

Fahrenheit	Celsius	How It Feels
113	45	Way Too Hot
86	30	Too Hot
59	15	Mild
32	0	Too Cold
5	−15	Much Too Cold
−22	−30	Not Recommended

Diameter	People	Size Name
8 in.	1	Small
12 in.	2	Medium
14 in.	3	Large
16 in.	4	Extra Large
133.2 ft.	40,000	World Record

Speed	Time	Mode of Transit
3 mph	20 min.	Walking
8 mph	7.5 min.	Running
15 mph	4 min.	Biking
30 mph	2 min.	Driving
120 mph	0.5 min.	Skydiving

If equations are complete but inaccessible and tables are accessible but incomplete, is there any way to combine the benefits of each? Is there a visual language that can convey a whole relationship at a single glance?

Hint: the title of this chapter.

Graphs work by the logic of latitude and longitude. When strolling the earth, you can describe your exact location using two numbers: a *latitude* (your distance north or south of the equator) and a *longitude* (your distance east or west of the prime meridian). This system turns geography into numbers: for every point on the planet, there's a unique pair of values.

Graphs are the same, but in reverse: for every pair of numbers, there's a unique point on the plane. They turn numbers into geography, and in the process, they create a picture more useful than any drawing of a bakery.

To begin, we draw an x-axis (our equivalent of the equator) and a y-axis (our equivalent of the prime meridian), meeting at a point called the *origin*.

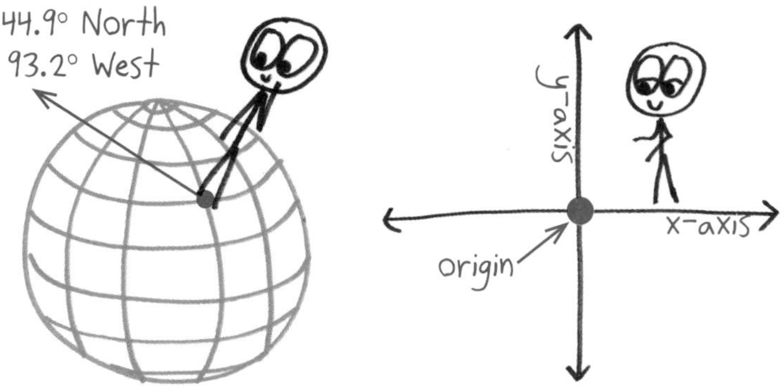

For any pair of numbers, the first one mentioned (x) specifies your distance to the right of the prime meridian, while the second (y) specifies your distance above the equator. Negative numbers signify the opposite directions—left of the meridian, and below the equator.

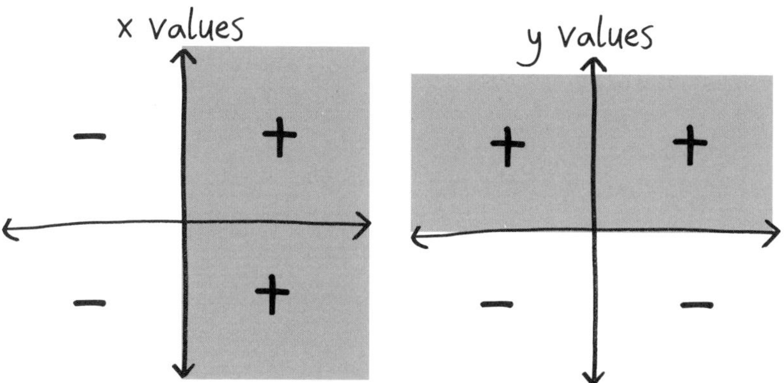

A pair of values thus becomes a point on the graph. For example, the equivalent temperatures of –12°C and 10.4°F become a single point: 12 steps to the left, and 10.4 steps up. We abbreviate this spot as (–12,10.4).

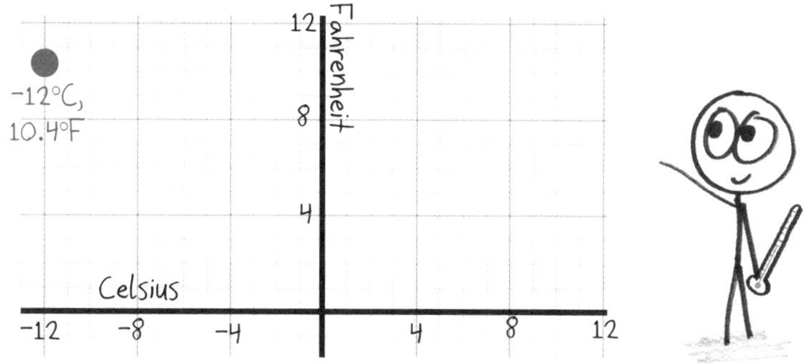

Several pairs from the same relationship create a necklace of points on the same graph.

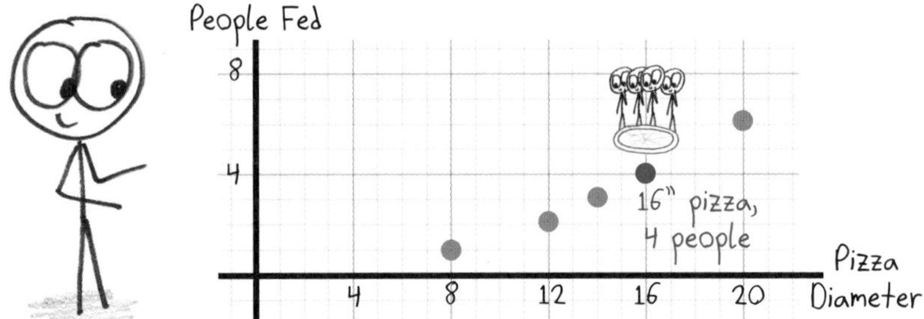

The whole relationship, with its infinite pairs, forms a continuous line or curve.

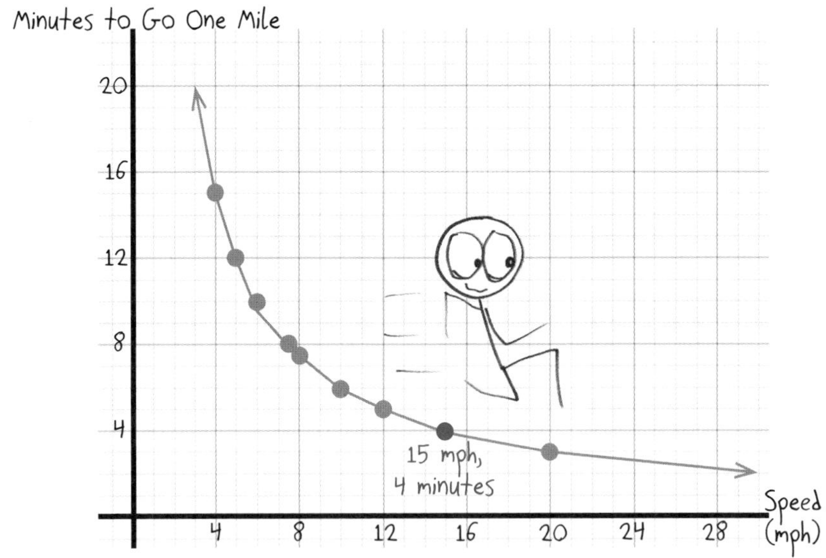

When I began teaching high school mathematics, I assigned 10, 15, or 20 graph problems a night. The students grumbled like workers at a badly managed graph factory.

The error was all mine: by having them graph and graph and graph and graph, without ever pausing to inspect or make use of their creations, I was prompting them to see graphs as final products, which they are decidedly not.

So what *are* graphs for?

For one, they indicate the scope of a relationship. Is zero a meaningful value? Are negatives acceptable? The graph tells us this information at a glance.

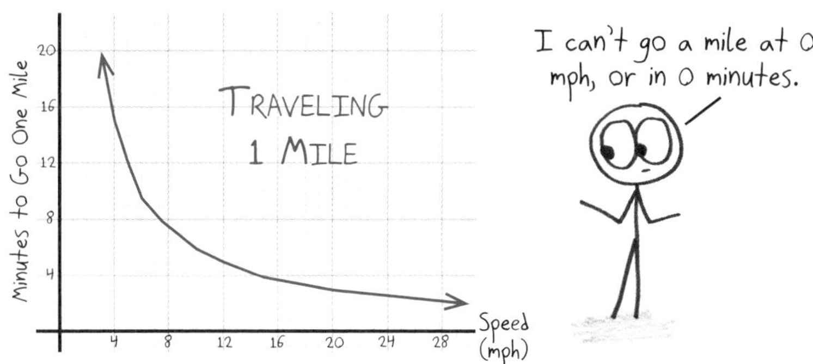

Graphs also reveal trends. As one number grows, does the other grow in proportion? Does it shoot off into the stratosphere? Or does it gradually approach a ceiling or a floor?

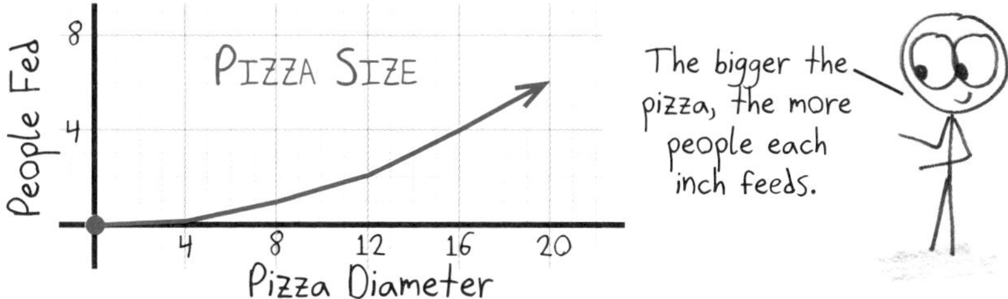

Graphs can even hint at special pairs of values. For example, is there a moment when both variables are equal? And what happens if one of the variables equals zero?

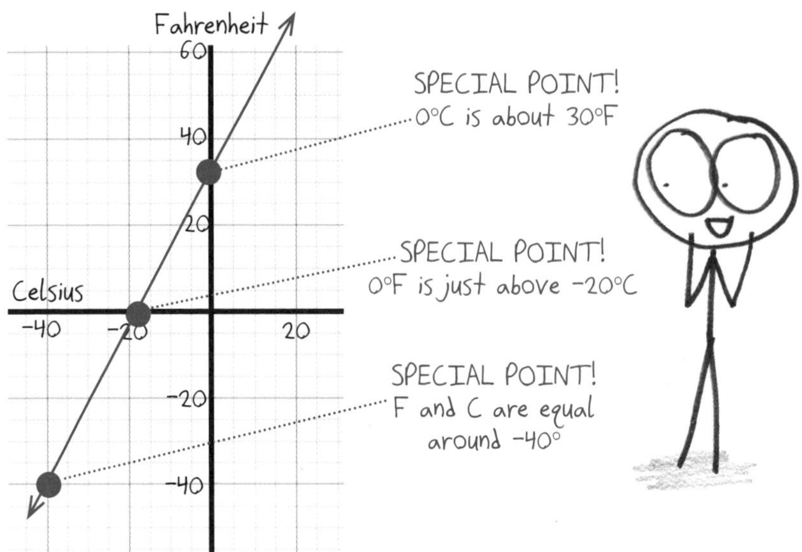

Graphs combine the completeness of equations with the readability of tables. But we pay a price: imprecision. For example, when $x = 7$, what exactly is y? Is it 8.5, 8.6, or 8.5714? An equation or table would tell us precisely, but graphs are too crude to distinguish nearby possibilities, just as a casual photo cannot reveal a person's height to the nearest millimeter. For that kind of exactitude, graphs aren't much good.

Instead, graphs are history's oldest form of data visualization: a method for turning a sprawling set of numbers into a single picture we can discern at a glance. "It is not how much information there is," said Edward Tufte, the godfather of data visualization, "but rather how effectively it is arranged." Graphs arrange a staggering amount of information (infinite pairs of values) into a singularly effective form. "Graphical excellence," Tufte wrote, "is that which gives the viewer the greatest number of ideas in the shortest time with the least ink in the smallest space."

The old maxim really is true: *when stuck on a math problem, draw a picture.* But to avoid the pitfalls of cheetahs and fighter jets and beautifully shaded bakeries, an addendum is necessary: *when stuck on a math problem, draw a picture that would make Edward Tufte proud.*

Formulas

The sci-fi writer John Scalzi has a crackpot theory that I love: every strawberry on Earth, no matter how large or small, has the same total amount of flavor. Big strawberries spread their flavor thin, with a faint trace in every bite. Tiny strawberries pack their flavor dense, with a wallop in every nibble. Call it the strawberry formula: *flavor intensity is inversely proportional to strawberry volume.*

I suspect it's not strictly accurate. But in a world of hulking apple-sized, weak-tasting strawberries, it sure *feels* accurate.

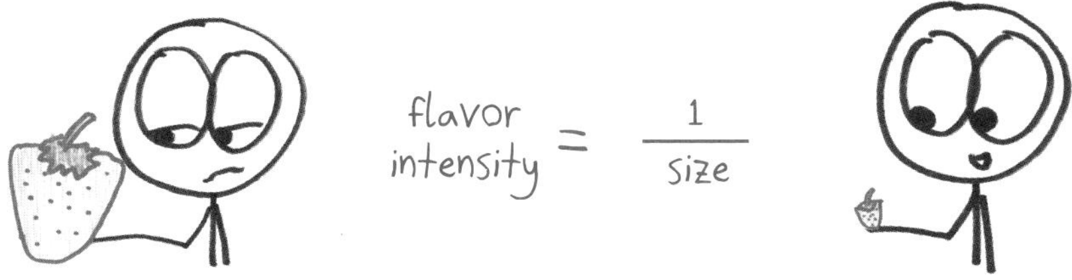

While I was outlining this book, my editor, Becky, asked if I planned to include a chapter on formulas. I had to ask for clarification. Did she mean the quadratic formula? The formula for the area of a trapezoid? The Scalzi strawberry formula?

"I just feel like school math was full of formulas," she said. "Formulas I had to memorize. Formulas we used over and over again."

I had forgotten how such equations loom over students. How you spend whole nights plugging numbers into $A^2 + B^2 = C^2$. How high school physics is a yearlong exercise in manipulating the same six or seven standard equations. How all these formulas seem to be fixed and eternal, like the chemical formula for water (H_2O), or the secret trade formula for Coca-Cola (REDACTED). I had forgotten, in short, how it feels to learn math.

I had forgotten all this because to a mathematician the whole "formula" concept is fuzzy. Formulas are just equations, and equations are just statements that two things are equal. There is no bright line separating $A = \pi r^2$ (the famous formula for the area of a circle) from $A = P^2 - 3d$ (the nonfamous formula I just made up). It's just that $A = \pi r^2$ is a lot more helpful.

A formula is nothing more than a wise equation.

So what counts as a formula? Not everyone agrees. Some textbooks include formulas that I'd dismiss as overly specific advice, less like "Be the change you wish to see in the world" and more like "Turn right on Randolph Avenue." Other formulas strike me as obvious common sense, less like "Don't skimp on your wedding photographer" (counsel I appreciated) and more like "Don't hit on your wedding photographer" (counsel I did not require).

Still other formulas are as arbitrary and fanciful as Scalzi's strawberry equation. For example, as a kid, I always wondered who classified books as "fifth-grade level" or "sixth-grade level." Some kind of reading expert? A handpicked focus group?

Nope. The publisher just plugged the text into the Flesch-Kincaid grade level formula:

words per sentence × 0.39
+ syllables per word × 11.8
− 15.59 = Grade Level

The formula pays no heed to what the words and sentences actually say. Short sentences full of short words? Low grade level. Lengthy, languorous sentences teeming with Latinate polysyllables? High grade level.

By this formula, the theoretical minimum reading level is −3.4. That would be a string of one-word sentences, each one syllable long. The closest you'll find is probably Dr. Seuss's *Green Eggs and Ham*, which has sentences of varying length but words almost exclusively of a single syllable. It scores −1.3, so I suppose you can read it a year before entering kindergarten.

At the other extreme, Lucy Ellmann's award-winning *Ducks, Newburyport* consists of a single sentence that lasts a thousand pages. By my estimate, the novel's Flesch-Kincaid score is over 150,000. That explains why no one I know is qualified to read it (except my friend Katy, a book critic who has apparently completed 35,000 bachelor's degrees).

Text	Difficulty
"Live. Laugh. Love."	Baby level
"I came. I saw. I conquered."	Kindergarten level
"Dance. Dance. Revolution."	Pretty difficult
"Ontogeny recapitulates phylogeny."	A thing my 9th-grade bio teacher Ms. Nulty told us, but which I needed four PhDs to comprehend

The Flesch-Kincaid grade level formula is a useful tool. It is also manifestly silly. It assigns equal difficulty to "the armadillo fell into the water" and "the legislature lapsed into rank bedlam." But I don't blame Flesch or Kincaid. No simple equation could ever capture the difficulty of all books, for the simple reason that no equation is perfect. The only universal truth is that no truth is universal.

Except in mathematics. You see, the formula $A = \pi r^2$ really does capture the area of all circles.

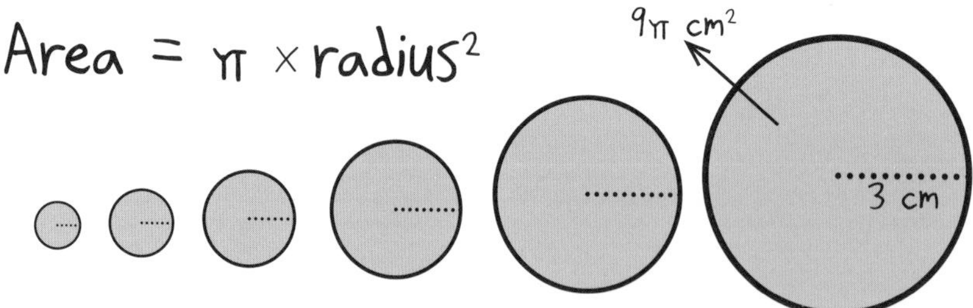

The formula $A^2 + B^2 = C^2$ truly does govern the sides of every right triangle.

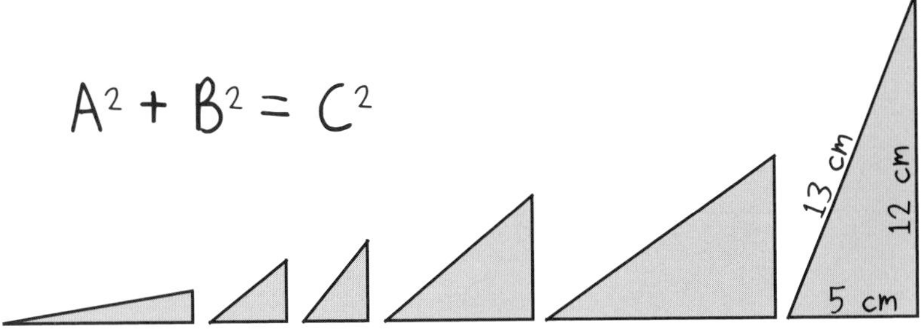

The formula $V + F = E + 2$ actually does take inventory of every cube, prism, and pyramid.

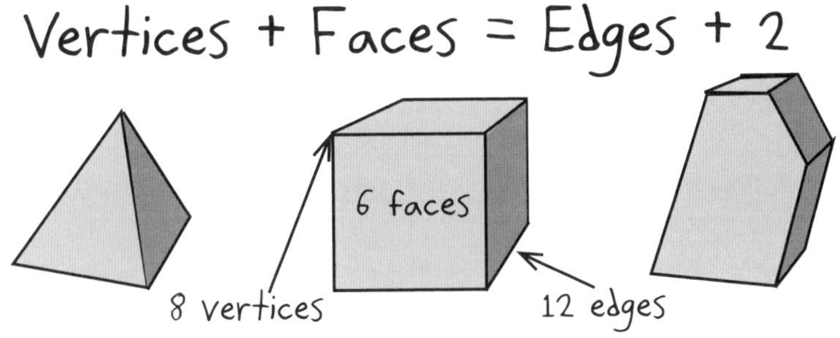

In real life, formulas have limited use. Sometimes they crop up as playful aphorisms. (It is said that the philosopher William James once defined "self-esteem" as "success divided by pretensions.") Other times they form rough-and-ready bureaucratic tools. (The Flesch-Kincaid formula was commissioned by the US Navy, to help them keep instruction manuals simple and readable.) But we all know that self-esteem, linguistic complexity, and strawberry intensity cannot be measured or calculated, except in the crudest sense.

Mathematics, suffice it to say, is not real life. It is populated by perfect abstractions, for which formulas are not simplistic approximations, but the actual truth, the whole truth. If my students look back on formulas as objects of mind-numbing repetition, then I have failed as a teacher. More than mere tools, formulas are the Great Books of mathematical literature, the canon of enduring equations.

Simplifying

One fall in college, I took a math class that was too hard for everyone. The professor didn't want to fail us all, so when finals rolled around, he gave a take-home exam on material that was much easier than anything we'd covered. The day we turned it in, my classmate Daniel asked the obvious question: "What did this have to do with the course?" The professor replied with an animated explanation lasting several minutes.

When he finished, Daniel summarized: "So... basically, nothing."

The professor smiled and shrugged, unable to dispute the paraphrase.

See, the relevance is this expression here.
That equals zero.
$\left(\frac{\sin x \cos x}{\sin 2x}\right)^{-2} - \left(\frac{\cos^2 x + \sin^2 x}{\cos(\frac{\pi}{8})}\right)^2$

Much mathematical work is just glorified paraphrasing. The world presents you with some kind of complicated expression; your job is to restate the information again and again, each time a little clearer and briefer than the time before. By this process, a several-hundred-word speech (*The relevance to our course is that the orientation-preserving symmetries of hyperbolic space permit an interpretation as...*) may become a single phrase (*basically, nothing*).

This kind of paraphrase is known as "simplifying." The word itself is controversial, as we'll see later, but the principle is hard to deny: simplicity leads to clarity, and clarity leads to insight. "Our life is frittered away by detail," wrote the naturalist Henry David Thoreau. "Simplify, simplify, simplify!"

To illustrate the principle, it helps to imagine an English-language equivalent. We begin with a doughy, bloated sentence, and then begin to fuss and fiddle with it.

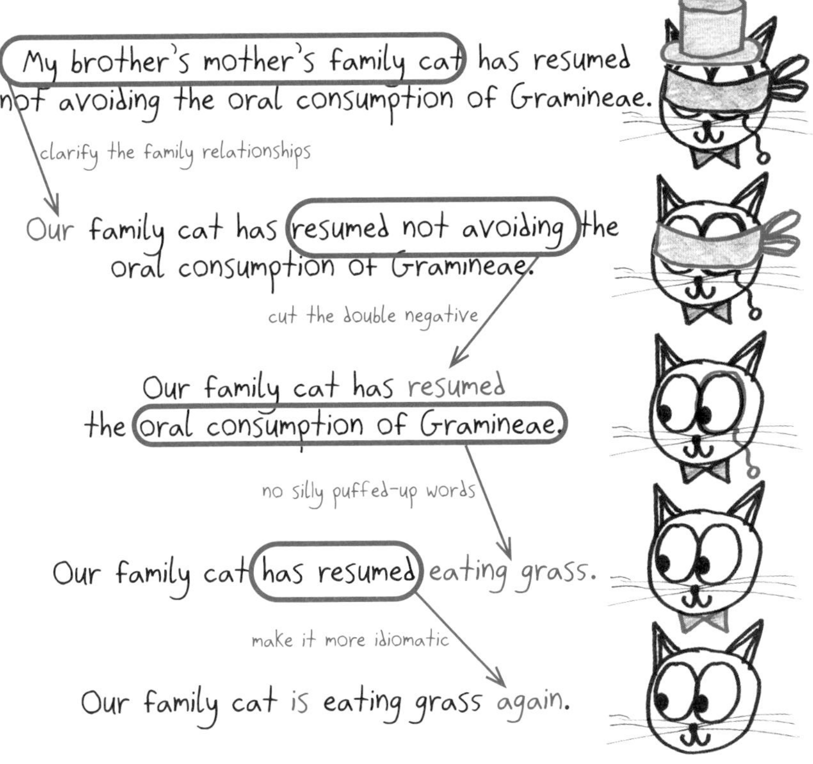

We did not alter the contents of the sentence in any way. Rather, we revealed them, by translating the message into a form better suited for human (or feline) consumption.

Admittedly, this was an artificial example. Such puffery and obfuscation are rarely seen in English, except as spoken by contract lawyers. But in math, such complications arise all the time. The world supplies information in a messy, piecemeal form, and only a judicious paraphrasing can render it into a more useful one.

For example, here is the thrust of one of my favorite trigonometric proofs. Don't sweat the details; just watch how a gnarly expression evolves into a final form of splendid concision.

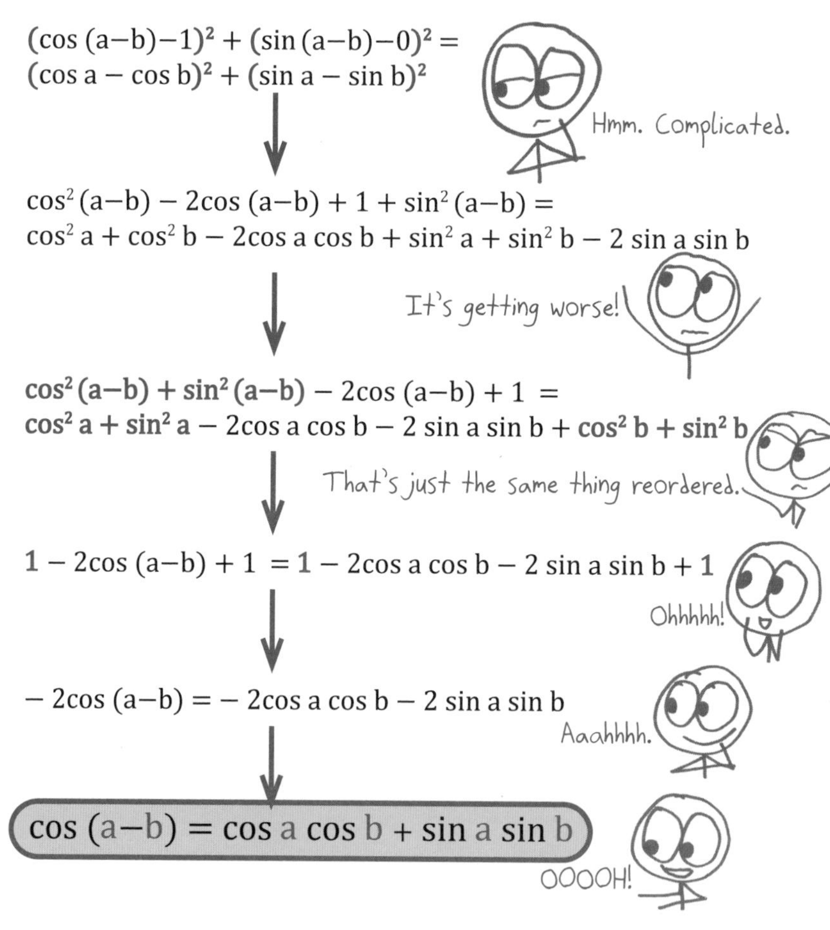
$(\cos(a-b)-1)^2 + (\sin(a-b)-0)^2 = (\cos a - \cos b)^2 + (\sin a - \sin b)^2$
Hmm. Complicated.
$\cos^2(a-b) - 2\cos(a-b) + 1 + \sin^2(a-b) = \cos^2 a + \cos^2 b - 2\cos a \cos b + \sin^2 a + \sin^2 b - 2\sin a \sin b$
It's getting worse!
$\cos^2(a-b) + \sin^2(a-b) - 2\cos(a-b) + 1 = \cos^2 a + \sin^2 a - 2\cos a \cos b - 2\sin a \sin b + \cos^2 b + \sin^2 b$
That's just the same thing reordered.
$1 - 2\cos(a-b) + 1 = 1 - 2\cos a \cos b - 2\sin a \sin b + 1$
Ohhhhh!
$-2\cos(a-b) = -2\cos a \cos b - 2\sin a \sin b$
Aaahhhh.
$\cos(a-b) = \cos a \cos b + \sin a \sin b$
OOOOH!

The work changed no information. It was a parade of paraphrases. Synonym by synonym, we nosed and noodled our way toward a more compact expression.

On display here is a deep principle of mathematical language: to clarify, we simplify. "The heart and soul of much mathematics," writes mathematician Barry Mazur, "consists of the fact that the 'same' object can be presented to us in different ways." A proof or calculation is often nothing more than a wise and witty change of language.

Consider the process of "solving" an equation. Visually, we seem to peel away layers of wrapping. But conceptually, we are only restating a fact. We might be told, for example, that two identical mystery boxes plus $4 are worth a total of $30.

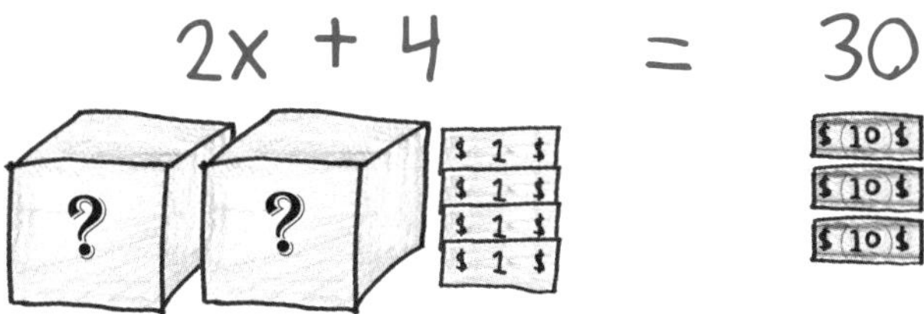

Setting aside the loose $4, the two boxes must be worth a combined $26.

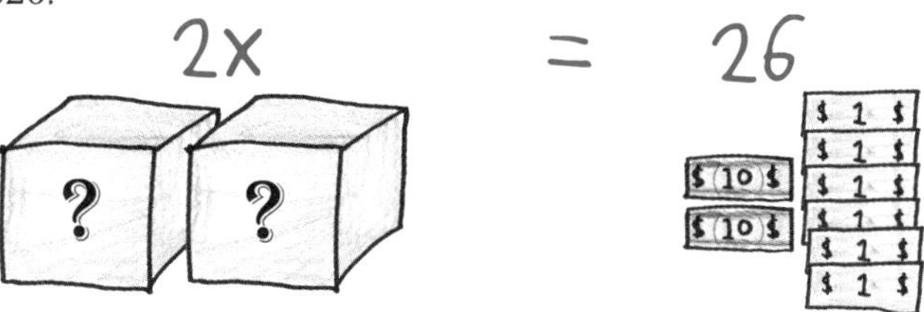

Now, to say that two identical boxes are worth $26 in total is to say that each one is worth $13. Equation solved.

Alas, what counts as "simpler" is not always clear-cut. Consider two equivalent expressions: $3x + 6$ and $3(x + 2)$. The first is a sum: three x's plus six 1's. The second is a product: three groups, each of size $x + 2$. Each phrasing is useful. Each is meaningful. But neither is inherently simpler than the other.

"The point of doing algebra," writes math teacher Paul Lockhart, "is...to move back and forth between several equivalent representations, depending on the situation at hand and depending on our taste. In this sense, all algebraic manipulation is psychological."

Old information, new formation.

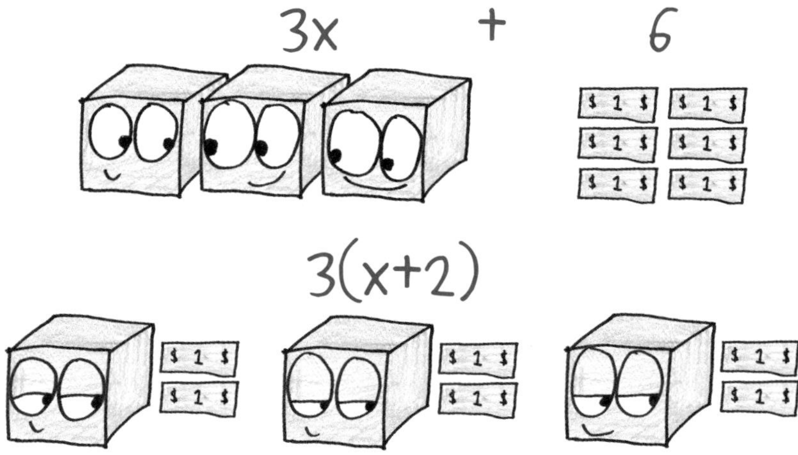

The ambiguity of the word "simplify" rankles some teachers. It is keenly frustrating when a textbook asks you to simplify $3x + 6$ into $3(x + 2)$, and then a few pages later to simplify $3(x + 2)$ right back into $3x + 6$. In such cases, a little jargon will achieve greater precision: instead of "simplify" and "simplify," we can say "factorize" and "distribute," or perhaps "condense" and "expand."

Still, when radicalized teachers take an extra step and propose to ban the word "simplify" entirely, I say they go too far. "Make everything as simple as possible," advised Einstein, "but no simpler." Yes, simplicity is a fuzzy ideal. But so are all the things we hold dear, from beauty to truth to the nature of a sandwich. Banning the word "simplify" is one simplification too many. Merely switching names does not clarify the underlying reality.

Except, I should say, in math itself. There, eliminating one phrase and replacing it with another really can bring a kind of enlightenment. In the language of math, and perhaps *only* in that language, wisdom can be as simple as a paraphrase.

Solutions

I enjoy whodunits, but I'm garbage at solving them. You could edit a mystery film so that from the very first moment, the murderer is labeled on-screen, and I'd still be flabbergasted by the grand reveal. "Ahh," I'd say. "The pieces are all coming together: the missing jewel, the forged signature, the blinking THIS IS THE MURDERER arrow. What an ingenious puzzle!"

Perhaps I should stick to mathematical whodunits. Though they differ in a few particulars, the premise is the same. We begin with a description of something (often an unknown number). Then, to "solve" this description is to figure out exactly what is being described.

For example:

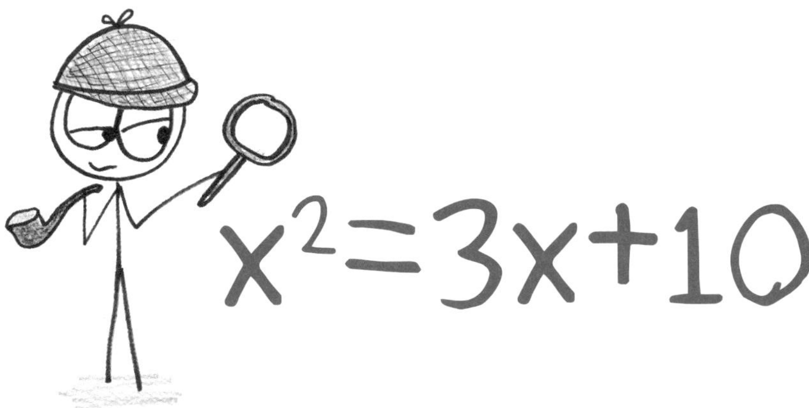

"There's a number known only as *x*," this equation says, "whose square is 10 more than its triple." This is not a question or a command, but a simple declaration of fact. Still, it stokes our curiosity. Who *is* this mysterious *x*? We become detectives of a kind, looking for a suspect that fits the description—in mathematical terms, a number that *satisfies* the equation. Our only clue is the equation itself.

In such a case, one approach is to guess at random. I don't recommend this method in criminal investigations, but in mathematical ones, guessing and checking often works, because the guesses can easily be checked.

For example, does the number 1 fit the description above? Its square is 1. Its triple is 3. Now, is the square 10 more than the triple? Alas, not even close. We cross 1 off our list. Okay, what about 2? Again, nope: its square (4) is smaller than its triple (6). How about 3? Still no: its square and triple are equal (both 9).

Keep going, and you may stumble across a solution: namely, 5. Its square (25) is precisely 10 more than its triple (15). Mystery solved.

Or is it?

Mathematics departs from Agatha Christie in that mathematical whodunits may not have a single solution. A description may have countless solutions, or no solution at all, or any number of solutions in between, like 17 or 5836.

ONE SOLUTION
7=x+1
Aha! The murderer!
INFINITE SOLUTIONS
3x=x+2x
We are *all* murderers...

ZERO SOLUTIONS
x=x+1
There is no murderer. In fact, the murder as described is impossible.

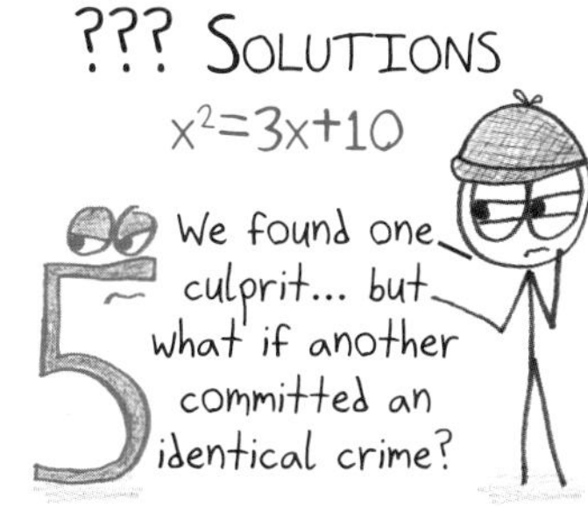

??? SOLUTIONS
x²=3x+10
5
We found one culprit... but what if another committed an identical crime?

A great deal of mathematics is, in effect, the study of various kinds of whodunits. You learn how many solutions to expect. You learn how to handle the evidence with care and caution. You learn the methods appropriate to each case. You grow into a kind of numerical Sherlock Holmes.

In this example, we can draw on techniques taught at the academy (don't worry about the details) to paraphrase our evidence in a more useful form:

$$(x - 5)(x + 2) = 0$$

The whodunit now says "Two numbers multiply together to give zero." This is the algebraic equivalent of a smoking gun. Heck, it's almost an open confession. When does zero emerge from multiplication? Only when one of the multipliers is itself zero.

If the first number $(x - 5)$ is zero, then x is 5. That's the solution we already uncovered.

Meanwhile, if the second number $(x + 2)$ is zero, then x is -2. That's a fresh solution, a murderer we had previously overlooked. Not an accomplice, exactly; more like a different criminal who happened to commit an identical crime.

The mystery is now solved.

Not every whodunit is so pleasing. In some cases, you may run into a cheap or uninteresting solution, like a mystery novel that's obvious from the second page. This is known as a *trivial* solution. For example, $x^2 + 2x = x^3$ is a pretty juicy whodunit: "A number's cube is equal to its square plus its double." However, before you get to any interesting solutions, you stumble into a boring one: $x = 0$. Zero's cube, square, and double are all zero, so the equation boils down to $0 + 0 = 0$. True, but dull.

In this way, a trivial solution is *logically* satisfying, but not *emotionally* satisfying. It satisfies your equation, but not your curiosity.

Still, a trivial solution *is* a genuine solution. More dangerous are *spurious* solutions: impostors or fakes created as byproducts in our search for the real solutions. For example, let's solve $\sqrt{x+2} + x = 0$. Approaching this mystery in the natural way (again, don't sweat the details) you will arrive at two solutions: 2 and −1.

However, only −1 is genuine. Try 2 in the original equation, and you'll wind up with the nonsense claim that $4 = 0$. This spurious 2 is no solution at all, but an industrial waste product we must filter out, using the original equation as a filter. To be clear, we committed no error. It's just that the pursuit of true solutions (like −1) may sometimes generate false ones (like 2).

Your cleverly laid trap captured this known fugitive. Also, this bewildered child.
Take the genuine solution into custody, and give my apologies to the spurious one.

Traditional whodunits and mathematical whodunits share a lot in common. We begin with a description (of a crime/calculation). We seek a culprit (the criminal/number). We handle evidence (fingerprints/the equation), ruling out the suspects with convincing alibis ("I was in France"/"I don't satisfy the equation"). We get to exercise our deductive faculties/embarrass Dr. Watson by calling his questions "elementary." Best of all, we arrive at solutions, each one as satisfying as a key sliding into a lock.

But wait. There's a twist. Isn't there always?

In math, numbers rarely act alone. We study webs of interrelated numbers. A formula such as the chemists' classic $PV = nRT$ spells out precisely how several variables (in this case, a particular gas's *P*ressure, *V*olume, *T*emperature, and *n*umber of molecules) relate to one another.

Thus, instead of a single equation describing a single culprit, we may face a whole system of equations, all describing an *Ocean's Eleven*–style team of criminal conspirators.

For example:

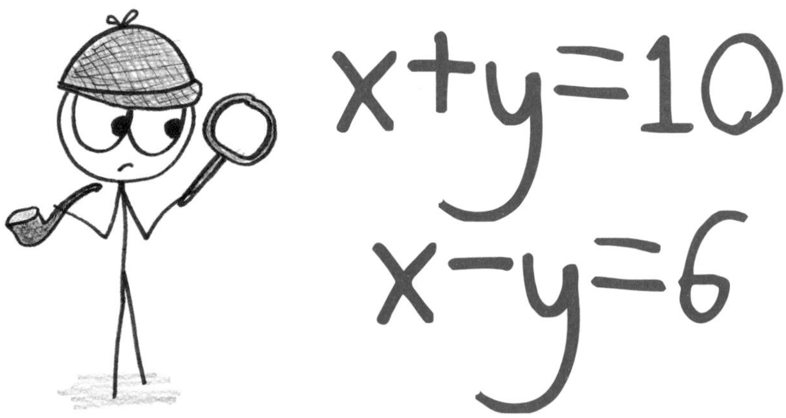

Here, we seek not one number, but a team of two. Fittingly enough, we have two pieces of evidence. First, their sum is 10, and second, their difference is six.

As it turns out, infinite pairs of numbers have a sum of 10, from 6 + 4 to 13.7 + −3.7. Similarly, infinite pairs have a difference of six, from 9 – 3 to 1006 – 1000. But the thrill of this mystery is that only one pair appears on both lists.

These are our culprits:

Like a detective who just cracked a case, I find myself boiling over with things to say about these sorts of multi-equation mysteries. They are the perfect culmination of algebraic language, the precise form of literature for which the grammar was crafted. Variables let us discuss unknown numbers; equations let us describe the relationships between them; and various clever simplifications allow us to determine the unknown numbers, in all their multiplicity and variety. You can even bring graphs into the mix, and watch as the whodunits blossom from written literature into visual art, a striking geometry of intersecting lines and surfaces...

But I leave those tales for other authors and thicker books.

When we began this journey, I claimed that the ideas of math are like a tree, and that the language is like a house built around it. I promised to take you inside the house; as of this page, I consider that promise fulfilled. But our work is not quite done. There are broader questions to answer, beginning with an unavoidable one: What should we do when mathematical language confuses and baffles us? How do we handle our own inevitable mistakes?

Category Errors

You learn a new language by making mistakes. Take my friend Roswell: His Spanish could always run circles around mine, in part because he was more eager to test his limits, more willing to stumble. When we arrived in Madrid for a high school exchange trip, he introduced me to his host: *"Esto es mi amigo Ben."* His host replied with an English-language scolding: "No, Roswell. We say *este* for a person, and *esto* for an object. You just said 'This thing over here is my friend Ben,' as if he were a rock."

Ah well. Speaking as a rock, no harm done. *Así se aprende.*

In learning algebra, one common species of mistake is the *category error*. A category error is not just wrong, like saying 3 + 4 makes 6. It is the wrong category of thing altogether, like saying 3 + 4 makes a low-sodium pumpkin pie. In one of his novels, Douglas Adams describes such an error as "a sort of mismatching of concepts, like the idea of the Suez crisis popping out for a bun." In his own novel, Lemony Snicket offers the example of a waiter who does not merely bring you the wrong food or drink but, also, bites you on the nose. To make a category error is to misapprehend not just a word or phrase, but a whole situation. Not just a part of speech, but the purpose for which we're speaking.

Here's a common category error. (Don't worry about what exactly the question means; we'll circle back to it.)

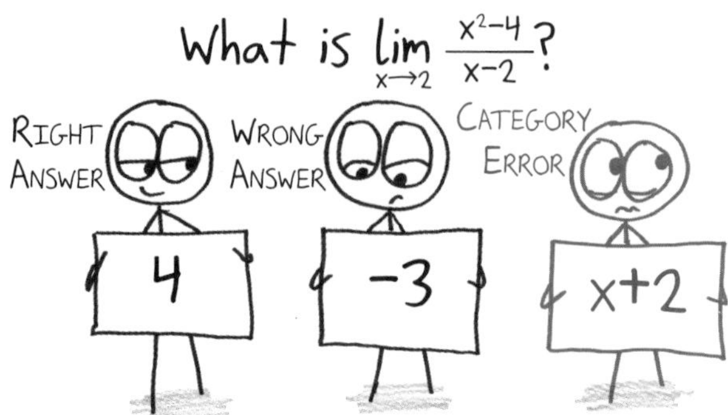

On the surface, the category error doesn't look so bad. In fact, on your way to the correct answer, the phrase $x + 2$ appears in the second-to-last line. "I only forgot the final step," a student could reasonably say. "It's not like I bit a patron on the nose. It's more like I forgot the slice of lemon in a glass of iced tea."

But taking an equivalent situation in English, the error is more stark:

The answer "3 a.m." is manifestly silly, but at least it's a time of day. The same can't be said for the category error, which gives the time only in relation to a mysterious, as yet unmentioned event. Two hours later than *what*?

The term "category mistake" was coined by the scholar Gilbert Ryle, in a book on the philosophy of the human mind. That's the kind of setting where such errors are prone to crop up: not in daily life, where the categories are clear and familiar, but in academic life, where the categories can be a bit airy and elusive. You wouldn't run into a coffee shop thinking it's a barbershop. But you might run into one abstraction thinking it's a different abstraction.

In mathematics, we have a habit of turning everything into nouns. An adjective ("three," as in "three blind mice") becomes a noun (3). A verb-like calculation (such as "multiply by four, then add two") becomes a noun expression ($4n + 2$). Whatever we encounter, be it a property or a process, we treat it as a *thing*, a noun.

So when everything under the sun is a noun, how can you tell which category of noun to utter?

Take our category error above. The question mentions a calculation $\frac{x^2 - 4}{x - 2}$, using the placeholder x to describe how the calculation is carried out. Then it asks, "If you run this calculation using numbers close to 2, you'll get results close to some other number. What is this number?" Phrased in those terms, it's clear the answer should be a number. While −3 is wrong, it's at least the right *kind* of thing. The answer $x + 2$ is the opposite: Though related to correctness, it is the wrong *kind* of thing entirely. I've asked for a number; you've described a calculation (adding 2).

It's quite analogous to the "What time does class start?" mistake. There, too, I asked for a number (4 p.m.) and you described a calculation (adding two hours). But in math, the mistake is far easier to make. You're only swapping one noun for another.

No, not THAT misshapen thing!
The OTHER misshapen thing!

A more frustrated and embittered version of me might have written this whole book as a catalogue of errors. Don't write $(a + b)^2$ as $a^2 + b^2$. Don't cancel the x's in $\frac{x}{x + y}$. Don't slouch, don't pick your nose in public, and never, *ever* divide by zero.

But why hector and fuss over honest mistakes? I prefer the wisdom of Roswell, which was perhaps best phrased by architect (and math lover) Piet Hein:

The road to wisdom?—Well, it's plain
and simple to express:
Err
and err
and err again
but less
and less
and less.

Category errors are communicative failures. They occur when something is lost in translation, when all of those nouns jumble together until you cannot tell one category from another. Who is a friend, and who is a rock? The only solution is to step back and ask, paying careful attention to questions of category and making sure to sift one sort of noun from another. *Así se aprende.*

Style

What makes mathematical language so hard to learn? We've already met a number of obstacles (or, if you will, an obstacle of numbers): the abstract nouns, the dense symbols, the verbs that aren't verbs. Yet beneath all these lurks another challenge: a looming, unspoken elephant-in-the-room sort of challenge.

We must separate the world being described from the language describing it.

In English, this is not a challenge. It is barely even a task. With no conscious effort, you separate a symbol (the letters *c-a-t*) from the thing it symbolizes (a whiskered animal). That separation is the essence of language: things have names, but the names are not the things.

In math, this line sometimes blurs. Let's take a moment to sharpen it.

For example, we generally write $\frac{1}{2}$ instead of $\frac{2}{4}$. But this is only a linguistic convention: a matter of form, not content. An English analogy might be passive vs. active voice: "I rowed the boat" vs. "The boat was rowed by me." Though the two carry the same meaning, most speakers favor the active verb. Likewise, when we favor $\frac{1}{2}$ over $\frac{2}{4}$, it's fundamentally a matter of style.

FACTUAL	STYLISTIC
What we say	How we say it
Content	Form
Who rowed the boat?	Passive vs. Active
What number is it?	Choice of denominator

Let me give another example. It's a fact that, when you multiply together two numbers, the result is not affected by order. Thus, $3 \times x$ is the same number as $x \times 3$.

Meanwhile, it's a convention that we never write $x3$, always $3x$.

It's not that $x3$ is inaccurate. Everyone will know what you mean. But it's an error nevertheless, akin to "two mouses" or "I gotted it." Cute, a bit distracting, and not entirely fluent.

In the same vein, the number $x \times y \times z$ can be written in six different ways, each corresponding to a different picture (see next page). But trying to juggle all six would be confusing; easier to pick one and make it the standard. So, as a matter of convention, we always write the variables in alphabetical order: not xzy or yzx, but xyz.

I like this rule because it's rarely even taught. Mathematicians apply it automatically, the way fluent English speakers always say "a big ugly bath toy" and not "a bath ugly big toy." In fact, as Mark Forsyth points out in *The Elements of Eloquence*, English adjectives are typically placed in a certain order: opinion, size, age, shape, color, origin, material, purpose. Hence, "a lovely little old rectangular green French silver whittling knife."

Likewise, in math, numbers are typically multiplied in a certain order: numeral, radical, constant, variable. Hence, we write $3\sqrt{2}\,\pi x y^2 z$, and never (unless we're trying to provoke someone) $zy^2\sqrt{2}\,\pi x3$.

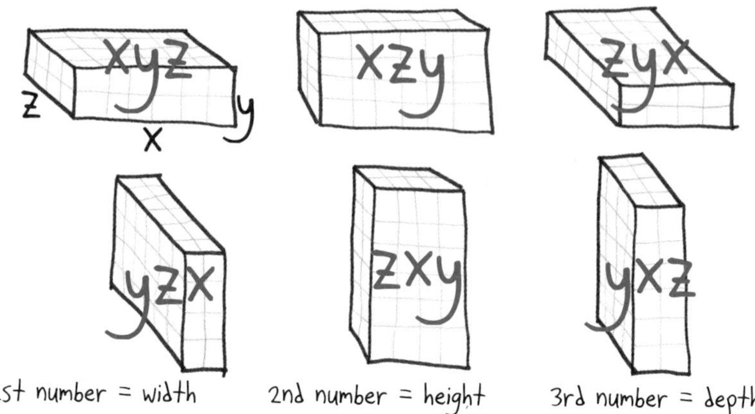

I could give more examples, but I hope the point is clear. Mathematical speech is not all a matter of right and wrong. Like English, it raises questions of style.

The difference is that in English, no one struggles to tell factual errors from linguistic ones. "Dogs can swims": true but ungrammatical. "Dogs cannot swim": grammatical but untrue. But if the distinction is so plain, then why do those learning math struggle to separate form from content?

It's tempting to blame educators like myself. We tend to give binary feedback: yes/no, true/false, right/wrong. Mistakes of content (writing $3x$ as $3 + x$) and of form (writing $3x$ as $x3$) incur the same verdict (*wrong*) and the same kind of penalty (*minus points*). It's as if parents were to respond with equal severity to bad grammar ("I loves you, Mom") and to bad meanings ("I detest you, Mother").

You could argue that we do this to impart the language at its purest and cleanest, but the effect is quite the opposite: by dishonoring the distinction between what you say and how you say it, we muddy the language's very nature.

Still, I don't blame the teachers. Nor do I blame the students. The blame, in my eyes, falls on mathematics itself.

The word "mathematics" refers to an invisible and intangible realm of ideas. It also refers to our language for this realm. It is both the world and the words for it, the symbols and the things they symbolize. Is it any wonder that we mistake the two? I can point to a cat and thereby teach you the word *cat*, but how can I point to a variable? How can I make you see that *x* and *n* are only names, that the words are not the world?

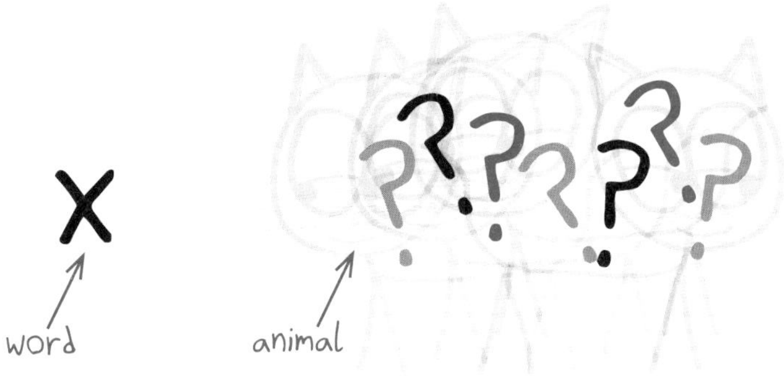

Rules

Toward the end of a sixth-grade lesson, a chipper young fellow named Kieran raised his hand. "I don't really understand anything you're saying," he informed me. "But I can still get the right answer." He beamed a patient smile.

I stifled a sigh. "Which part can I help you with?"

"Oh, I don't need help," he said. "It's just that you were talking about this extra stuff. Like, the ideas behind it. I don't, you know, *do* that."

I blinked. He blinked. A great silence passed between us.

"Is that okay?" he concluded. "I mean, as long as I can get the right answer?"

There it was, out in the open: the subtext of almost every lesson I had taught that year. Day after day, I tried to illuminate the logic behind the symbols. Day after day, my students politely ignored my prattle to focus on the symbols themselves. What made that afternoon stand out was that Kieran broke the fourth wall. He uttered the title of the film we were acting in.

To do math, must you think about the ideas, or can you just focus on the symbols?

On this particular day, we were exploring a rule from our chapter on multiplication: the distributive property. It's a logical fact about breaking apart larger piles into smaller ones. For example, with piles of 17, you can break each one into a pile of 10 and a pile of 7. Or, more generally, with piles of $b + c$, you can break each one into a pile of b and a pile of c.

We can distill this general truth into a compact symbolic form, like this:

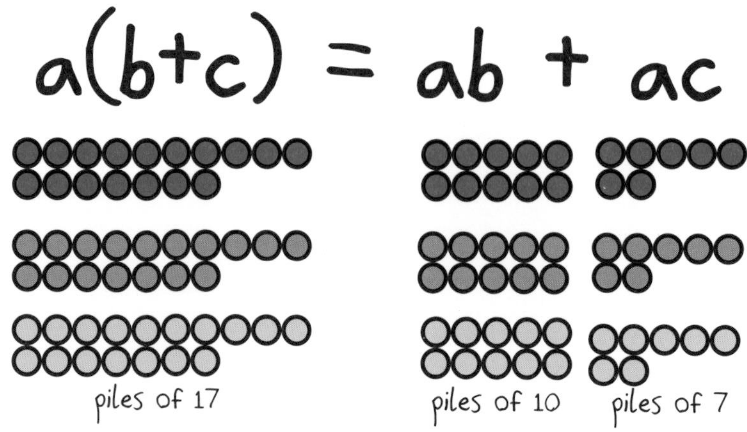

Alas, compact symbolic forms are more dangerous than they sound. Before long, students stop seeing $a(b + c) = ab + ac$ as a deep truth about multiplication, addition, and the logic of piles. Instead, they begin seeing it as a rule about letters and parentheses, a convention for how symbols move across a page. They conclude that text in the form $a(b + c)$ can be replaced with the text $ab + ac$, no matter what the marks mean.

That's when they start dashing off confident pronouncements such as these:

$$\log(b+c) = \log(b) + \log(c)$$

$$\sqrt{b+c} = \sqrt{b} + \sqrt{c}$$

$$(b+c)^2 = b^2 + c^2$$

Which all look fabulous, except that if you try some numbers, none of them are true.

0.301 0 0

$$\log(1+1) \neq \log(1) + \log(1)$$

5 3 4

$$\sqrt{9+16} \neq \sqrt{9} + \sqrt{16}$$

25 4 9

$$(2+3)^2 \neq 2^2 + 3^2$$

With great labor, students learn to recognize these as errors. But rarely do they grasp why. They seem to prefer memorizing each fact as an additional rule about how symbols move, a set of intricate exceptions to the initial $a(b + c) = ab + ac$ rule.

Teachers call this approach *symbol pushing*. Just push the symbols around the page, and don't worry what they symbolize. It is a mechanical view of mathematics, a way to speak, but with no idea of what you're saying. As David Hilbert quipped, "Mathematics is a game played according to certain simple rules with meaningless marks on paper." That's symbol pushing in a nutshell. Language divorced from meaning. Enough to make any teacher sigh.

But a few weeks after Kieran's query, I learned that not all mathematicians share my dim view of symbol pushing. I mentioned to my dad (a mathematician himself) that I had started writing an essay titled "How to Avoid Thinking in Math Class." Before I could say any more, he gave the project his stamp of approval. "Great," he said. "I've always said that the point of math education is to help you not to think."

I was taken aback. No, I explained, the title was ironic. On the question of "Should we think?" I was firmly in favor.

"Oh, yes, thinking is good," he generously conceded. "But it's too hard to do all the time."

He (and, indeed, Kieran) had a point. For example, it is an algebraic truth that $(x + 1)(x - 1)$ is the same as $x^2 - 1$. This fact boils down to a repeated application of the distributive property; as such, you can explain it entirely in terms of pile rearrangement. But trying to do so is like ascending a sheer cliff face.

$(x+1)(x-1)$

"We have a number of groups, each having two more items than there are groups (for example, 9 groups of 11 items each). Now, if we remove a single item from each group, and gather these together, then we'll have shrunk each group by one item, so that now each contains only one more item than the number of such groups, while also creating an additional group, one item smaller than the rest. This means we are one item short of having the number of groups and the number of items per group be equal."

$x^2 - 1$

Quite a climb! Invigorating as an occasional workout, but unthinkable as a morning commute. This was exactly my dad's point: *Thinking is good. But it's too hard to do all the time.*

"Operations of thought," wrote the mathematician Alfred North Whitehead, "are like cavalry charges in battle—they are strictly limited in number, they require fresh horses, and must only be made in decisive moments."

In this case, there is no need to send in the cavalry. The symbol pusher, thinking only about letters and parentheses, reaches the same summit in a few effortless strides.

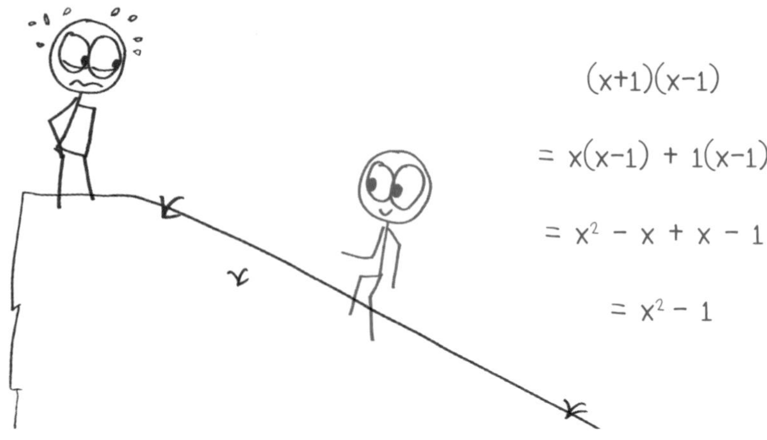

Symbol pushing isn't a cheat or a hack. It's a design principle. "I didn't recognize how I'd begun to reposition myself," writes Karen Olsson, of a youthful encounter with symbols, "how ready I was to disappear into a piece of paper—how the representation of a thing could seem more alluring than the thing itself."

Throughout the history of math, notations often gain popularity precisely because they lend themselves to simple mechanical rules. You could say we choose the symbols for the express purpose of pushing them around. We can then generate right answers with no insight, no inspiration, no input other than elbow grease. Just turn the crank, and new knowledge pops out.

Can you imagine if English worked this way? An object's name would indicate its physical size, so that a chihuahua (nine letters) would be three times the size of a cow (three letters). A food's name would encode its recipe, so that a pizza would be a "DoughSauce-CheeseBake." Chemistry would be a tediously safe area of study, because we could run experiments simply by smushing together the names of various chemicals and seeing which ones spell "explosion."

Symbol pushing boils the laws of logic down to laws of grammar. The language becomes a scale model of reality. We can wrangle ideas simply by wrangling ink.

So who was right, me or Kieran? The answer, of course, is both. To speak mathematics is to slip back and forth between two worlds, to inhabit two distinct frames of mind: the hard joy of thinking and the mindless trance of symbol pushing. Without the ink, the ideas are befuddling; but without the ideas, the ink means nothing. Learn the logic, learn it well, and then turn off your brain and let the symbols on the page dance to the silent music of the mind.

10⁻¹ 10⁰ 10¹ 10² 10³
0.1 1 10 100 1,000
π ≈ 3.14
τ ≈ 6.28
y=mx+b
Ah, simple.
around
across
"857"
1/4 + 1/6 =
-(-(-(-1)))
$2.5
125
+ 124
249
∀ε>0,
∃δ>0:
|x−a|<δ→
|f(x)−L|<ε
ax²
+++
THE FUZZY TRUTH
96°F
97°F
98.6°F
100°
unreasonable rounding
CHOCODOLLAR
10×10
1/10×10×10
3.14 π 22/7
3.140 3.141 3.142 3.143
17 × whatever
1.414ish
PI DAY PARTY!
3/14
3.4104398
+ 2.5899496
6.0003894
5×3
multiplication
4/7
10
farther
closer
2573
× 389
division by 1/2
17
34
division by 2
3 + −1.78
???
0 5 10 15 20 25 30 35 40 45 50
repeated addition
5⁵⁵
repeated exponentiation
5↑↑3
tetration
λ
4^(7/2) = ???
6
2
NOOOOOO
a÷b/c =
whisper conversation blender nightclub thunder
0 30 40 50 60 70 80 90 100 110 120
×1000
x(x+x²)?
x²+x³!
x²=3x+10
8÷2(2+2)
8÷2(4)
8÷8
1
K=½mv²
If A<B and B<C
then A<C
x = −b ± √b² − 4ac / 2a
DONUTS
7 PIZZAS
10/6
44.9° North
93.2° West
ax² = bx + c
ax² + bx = c
ax² + c = bx
ax² = bx
ax² = c
bx = c
1/3 sleep deprivation,
2/3 game theory.
Guaranteed 100%
Almost certain 95% to 99.9%
Very likely 80% to 95%
Probable 60% to 80%
Maybe 40% to 60%
Possible 20% to 40%
Unlikely 5% to 20%
Almost impossible 0.1% to 5%
Impossible 0%
AROUND ≈ 3 × ACROSS
OOOOH!
17/34
noisy
11
"1000"
1 −1 −(−1) −(−(−1))
5↑↑3 WOW!
30×20 30×7 8×20 8×7
600 + 210 + 160 + 56 = 1026
5+3 5×3 5³
No! Silence your nonsense!
82,771
What is lim x→2 (x²−4)/(x−2)?
0 .1 .2 .3 .4 .5 .6 .7 .8 .9 1
What's 46 + 28?
Let's see... calculating... calculating...
27

PHRASE BOOK

A Local's Guide to Mathematical Vocabulary

Although this book is almost done, I have not yet taught you the entirety of mathematical language. For example, we skipped matrix multiplication. Also matrix transposes, matrix groups, *Matrix* puns, and indeed, the entire concept of a matrix. We similarly missed differential calculus, integral calculus, nonstandard calculus, renal calculus, and—it is just now occurring to me—geometry. Like, all of geometry. It seems that, to the nearest percentage point, I have taught you 0% of mathematics.

Worry not, my friend. Such was the plan. This book is not an encyclopedia of the mathematical world, but only a brief introduction to the language for navigating it. Together we have stood on the shore and built a little boat; I leave the charting of the oceans to you.

Still, before you sail onward, I have one parting gift, more precious than any rudder or compass: a guide to our inside jokes.

If you have ever met a mathematician, you have perhaps noticed our peculiar manner of speech. Even our most banal conversations are sweetened with academic jargon. On the approach of spring: "It's not monotonic, but the weather is definitely improving." Or, comparing two restaurants: "I prefer the new one, but the menu is higher-variance." Or, after getting lost in a furniture store: "I just don't understand the topology of Ikeas."

The language of math developed for the purpose of naming abstract relationships. This makes it both very precise and very general. It makes crisp technical distinctions, yet it applies to almost anything, even situations with no identifiable numbers or shapes.

As such, our phrases sometimes slip into the popular lexicon. For example, "exponential" has been widely adopted to mean "growing really fast." A crude usage, but it captures the gist of the term. Similarly, "inflection point" has come to mean "the moment when a trend really took off" (which is almost the opposite of its mathematical meaning, but ah well). There is even a rising star: "orthogonal." The technical meaning is "perpendicular"; a rough synonym is "unrelated." In any case, it drew oohs and aahs from US Supreme Court justices when a lawyer used the term in oral argument.

"It's a cycle," notes Michael Pershan. "Math gobbles up language, assigns it new meaning, and then spits it back into popular usage."

In this final section of the book, I invite you to join the cycle. To explain each term in its precise mathematical context would require a whole stack of books. But to capture the flavor—to bring you within the radius of the inside joke—well, I believe we can accomplish that with just a few dozen cartoons, organized into a few loose topics.

Growth and Change

As they say, change is the only constant. Seasons turn, empires fall, and children outgrow their wardrobes every few hours. Who can furnish a vocabulary for this world of flux and flow? The philosophers? The poets? Oh, please; they're just as frightened and confused as we are. If we want to describe these shifting sands with any kind of precision, there's only one place we can turn: the icy clarity of mathematics.

delta—*noun*. The change or difference between two things.

jump discontinuity—*noun*. A sudden leap or change that skips past the middle steps.

chaotic—*adjective*. Highly unpredictable, so that nearly identical starting points may lead to dramatically different results. (Not synonymous with the standard English usage.)

exponential—*adjective.* Growing fast; more precisely, doubling whenever a fixed interval of time elapses.

derivative—*noun.* The rate at which something is changing: can be positive (the thing is increasing), negative (the thing is decreasing), or zero (the thing is not changing).

Errors and Estimates

Mathematicians don't get everything right. If you've ever seen me walk directly into a parking meter, you may suspect that mathematicians get *nothing* right. The truth is somewhere in between: mathematicians are as error-prone as anybody, but they have a knack for knowing when errors are most likely and for distinguishing big errors from small ones. The mathematician's secret skill, you might say, is their rich and evocative vocabulary of wrongness.

epsilon—*noun*. A tiny change or difference.

Okay, we still haven't found the hotel, but I believe we're epsilon closer than we were an hour ago!

first-order approximation—*noun*. A helpful starting point; a crude but useful first step in understanding something.

sign error—*noun*. The mistake of switching positive for negative, or vice versa.

point estimate—*noun*. A best guess.

confidence interval—*noun*. A range of values with a stated probability (say, 90%) of including the true value.

handwave—*verb*. To leave out the rigorous technical details and just capture the gist.

Optimization

The tragic condition of the 21st-century human is that we are always trying to make better things: Faster travel. Tastier dinners. Cuter dogs. Ours is a life of perpetual striving, which is to say, perpetual dissatisfaction. It's no way to live—but if we must, then we owe it to ourselves to speak about it with clarity and intelligence.

optimize—*verb*. To make some target variable as large (or small) as possible.

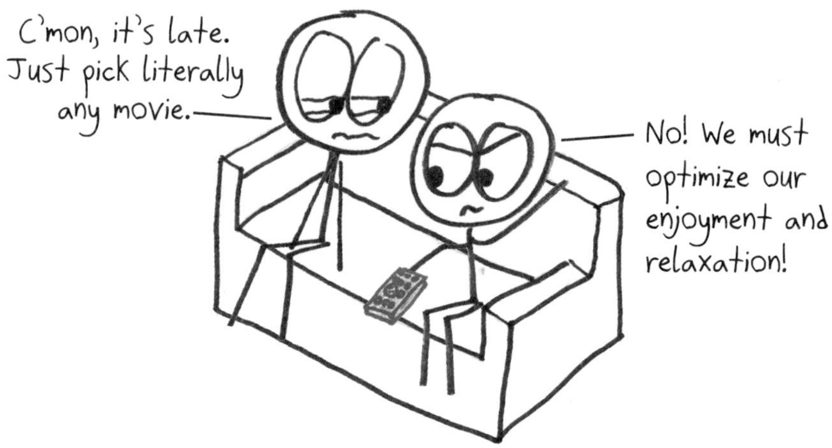

objective function—*noun*. The thing you are trying to maximize or minimize.

constraint—*noun*. A limitation on the options under consideration.

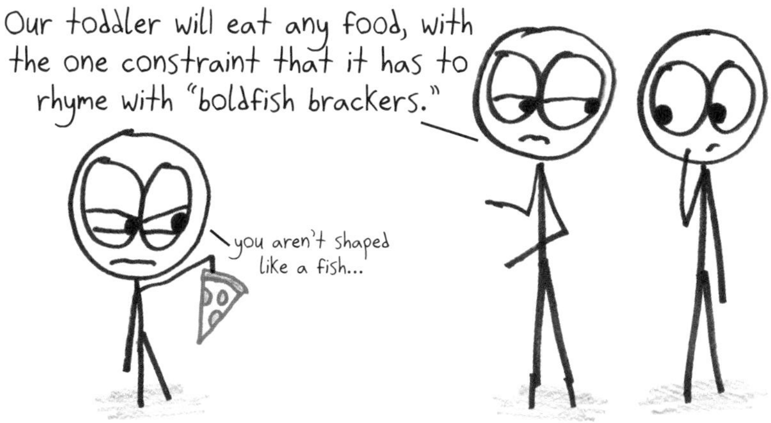

overdetermined—*adjective*. Having so many constraints that no solution is possible.

global optimum—*noun*. The very best we can do; the highest mountaintop.

local optimum—*noun*. The best we can do without dramatic changes; a hilltop, though not necessarily the highest one.

gradient descent—*noun*. The process of seeking a local optimum by making a series of tiny improvements.

Solutions and Methods

All of us—engineers, pastry chefs, stay-at-home parents—are in the business of solving problems. But with such different challenges—collapsed bridges, collapsed soufflés, collapsing toddlers—it can be hard to find a common language. That's where mathematicians come in. What we need is not their problem-solving prowess, but their nuanced language for describing problems—and, better yet, solutions.

algorithm—*noun*. A systematic way of solving a particular kind of problem.

heuristic—*noun*. A method that is quick and useful, though usually imperfect.

brute-force—*verb*. To solve a problem by systematically trying every single possibility.

elegant—*adjective*. Simple yet highly effective; the opposite of brute-force.

inverse problem—*noun*. For a given effect, the challenge of determining its cause.

Shapes and Curves

This book has focused on algebra, to the neglect of a whole different side of mathematics: *geometry*. The first grows from numerical thinking, and the second from spatial thinking. The great miracle of mathematics is that algebra and geometry converge, creating a kind of impossible tree with a single trunk yet two systems of roots. Numbers can be understood in spatial terms, and space can be dissected through numbers. In any case, some geometric concepts are too useful (and too mind-bending) to exclude.

high-dimensional—*adjective*. Having lots of aspects to consider.

geodesic—*noun*. Shortest path between two points. Not necessarily a straight line.

non-Euclidean—*adjective*. Not obeying the familiar rules of geometry.

Möbius strip—*noun*. A surface with only one side, so that its front and back are one; can be created by twisting a strip of paper and taping the ends together.

nonlinear—*adjective*. Not obeying a simple ratio; changing at different rates at different times.

Infinity

Not many mathematical ideas capture the popular imagination. Yet one concept has become a crossover hit, embraced as a part of theologians' doctrines, poets' rhetoric, and children's taunts: *infinity*. To mathematicians, infinity is more than just a single idea; it has different uses, different meanings, even different sizes. "Infinity" isn't just a word; it's a whole vocabulary.

unbounded above—*adjective*. Without a ceiling; able to get higher and higher and higher.

unbounded below—*adjective*. Without a floor; able to get lower and lower and lower.

dense—*adjective*. Pervasive; to be found in every nook and cranny.

countably many—*adjective*. In the mathematical hierarchy of infinities, the smallest kind of infinite.

uncountably many—*adjective*. A larger kind of infinite.

asymptotically—*adverb*. As eternity unfolds. (Used to describe an outcome that we don't necessarily reach, but to which we come progressively closer.)

Collections

Just about everything is made of sets. The US is a set of 50 states; *Abbey Road* is a set of 17 songs; and July is a set of 31 days. Thus, even the simplest act—say, building a Beatles playlist for your summer road trip across the United States—is secretly fraught with the mathematics of *set theory*. Lucky for us, mathematicians have developed a crisp and powerful language for describing such collections of things.

empty set—*noun*. A collection with no members at all.

I'm not picky. I just want someone emotionally stable, reasonably attractive, and with no annoying habits.
Ah, so you'll date anyone in the empty set.

subset—*noun*. A smaller collection contained in a larger one.

union—*noun*. Two collections combined.

intersection—*noun*. The members belonging to two collections simultaneously.

disjoint—*adjective*. Totally separate; nonoverlapping.

permutation—*noun*. A way of arranging or recombining elements.

closed under an operation—*adjective phrase*. Of a set: if two members are combined via the operation, then the result will also be a member of the set.

Logic and Proof

A product unique to the mathematician is the *theorem*: a statement that has been proved true beyond all doubt. Proving theorems is the mathematician's raison d'être: indeed, a mathematician has been described (by Alfréd Rényi) as "a machine for turning coffee into theorems." It's an even better line in German, where the word *Satz* means both "theorem" and "coffee residue"; thus, a mathematician turns coffee into coffee residue. And if you want to share coffee with a mathematician, it's vital to know the language of proof.

axiom—*noun*. A foundational assumption; almost an article of faith.

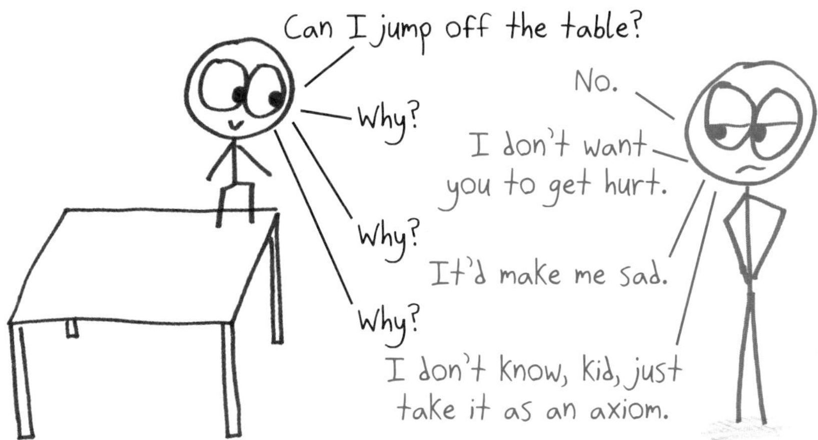

conjecture—*noun*. A proposed statement that may be true but hasn't been verified.

counterexample—*noun*. An exception that disproves a proposed rule.

theorem—*noun*. A rule that has been proved true.

constructive—*adjective*. Containing instructions for precisely how to create or find something.

existence theorem—*noun*. A proof establishing that something exists, but not where it is or how to find it.

Somewhere out there is the perfect person for you.
Okay, but where?

corollary—*noun*. A fact that is an obvious consequence of another fact.

I never want to see you again.
So... does that mean you won't be giving me a ride to the airport tomorrow?
Yes, that would be a natural corollary.

QED—*interjection*. A dramatic pronouncement to conclude an irrefutable argument. Abbreviation for the Latin *quod erat demonstrandum*, meaning "which was to be demonstrated."

Truths and Contradictions

My favorite definition of "mathematics" comes from mathematician Eugenia Cheng, who said that math is the study, using the rules of logic, of whatever will follow those rules. It rings true to me. The real essence of math isn't numbers, operations, shapes, or equations. It's logical reasoning: the study not of what's true, but of how various possible truths connect. If this sounds a little abstract and airy, then, yes, it is—which is precisely what makes the terminology so useful across all walks of life.

proof by contradiction—*noun*. An argument that temporarily assumes the opposite of what you want to prove, thereby revealing that the opposite is hopelessly wrong.

paradox—*noun*. An apparent contradiction, when two statements appear true but are logically irreconcilable.

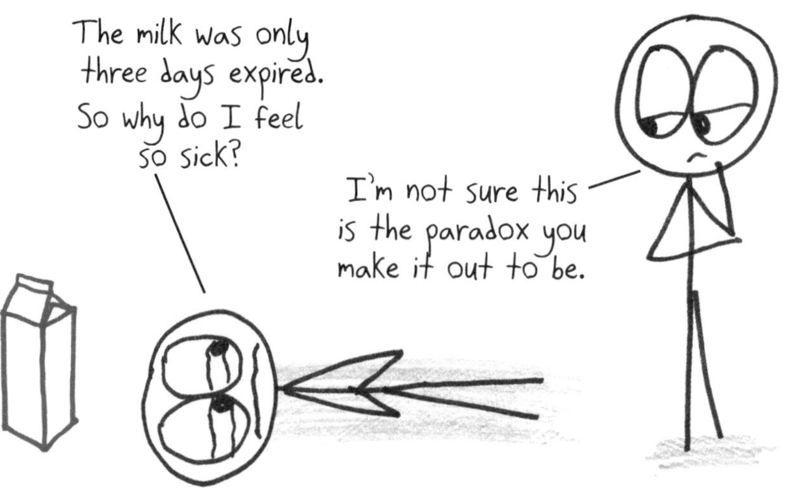

tautology—*noun*. A self-evident statement; something true by definition.

stronger—*adjective*. More sweeping than another statement; containing all of the other statement's implications and then some.

without loss of generality—*adverb*. In effect, "I will now discuss a specific scenario, but what I say applies equally to all scenarios."

special case—*noun*. A specific example of a general rule; may have some idiosyncrasies but should ultimately follow the broader pattern.

generalize—*verb*. To apply (or be applicable) to a broader context.

arbitrary—*adjective*. Not yet determined or specified; can also mean "generic."

Would you like to try a new restaurant tonight?

Like, an arbitrary new restaurant?

MENU

Actually, I meant the Filipino place.

The Probable and the Possible

Ben Franklin once quipped that nothing is certain except death and taxes. These days, with eccentric billionaires spending small fortunes to forestall death and large fortunes to forestall taxes, even those seem doubtful. In a world with no certainty, what do we do? Simple: We sort the likely from the unlikely, the probable from the improbable. We learn to distinguish the hundred meanings of "possible" and the thousand shades of "maybe." In short: we learn to speak *probability*.

probability—*noun*. Likelihood.

A PROBABILITY-TO-ENGLISH DICTIONARY

Guaranteed	100%
Almost certain	95% to 99.9%
Very likely	80% to 95%
Probable	60% to 80%
Maybe	40% to 60%
Possible	20% to 40%
Unlikely	5% to 20%
Almost impossible	0.1% to 5%
Impossible	0%

probability zero—*adjective*. Technically possible, but never going to happen.

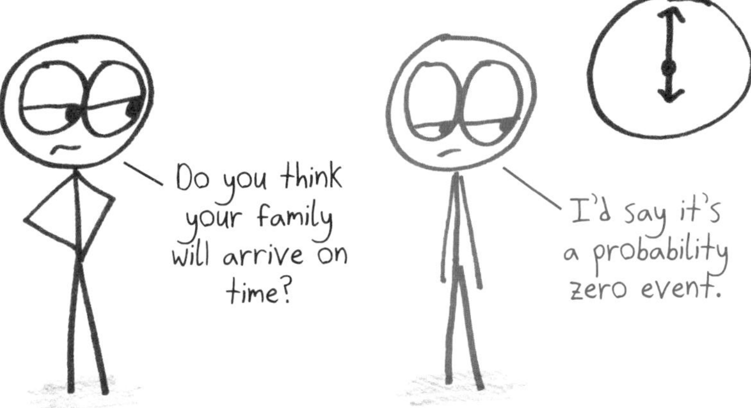

Bayesian prior—*noun*. Whatever you happen to believe before gathering any information.

update—*verb*. To modify your belief based on new information.

No, Madame Ghost, I still don't think this house is haunted. But I confess that our conversation has made me update my priors.

stochastic—*adjective*. A fancy synonym for "random."

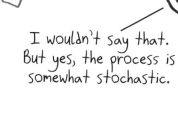

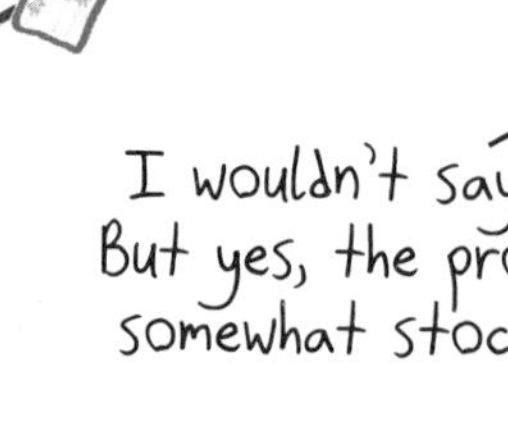

So your movie rating site just gives random scores?

I wouldn't say that. But yes, the process is somewhat stochastic.

conditioned on—*adjective.* Momentarily taking something for granted or assuming it to be true, even if it isn't.

Causes and Correlations

Life's deepest mysteries all boil down to a single word: *why*. Why are chocolate croissants so flaky and perfect? Yet why do I feel sick after eating a mere three or four of them? Do the croissants make me ill? Does illness drive me to seek out croissants? Mathematics may not have all the answers, but it has the perfect language for framing the questions: a language that separates *correlation* (two things going together) from *causation* (one thing causing another).

correlated—*adjective*. Of two variables: when one is bigger (or smaller) than average, the other tends to follow suit.

proportional—*adjective*. Of two variables: perfectly correlated, so that if one doubles, the other doubles, too.

negatively correlated—*adjective*. Of two variables: as one gets bigger, the other gets smaller.

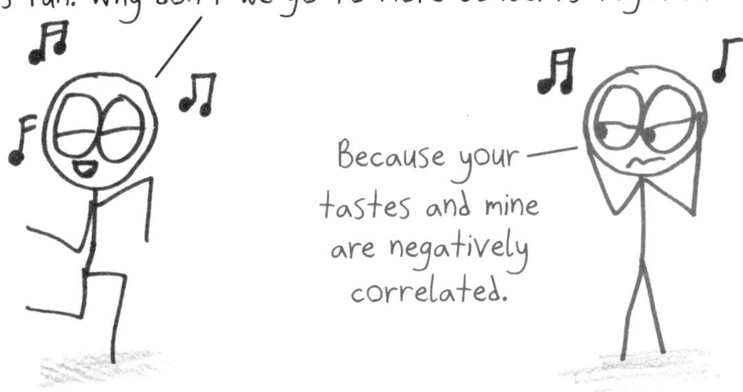

zero correlation—*noun*. A total lack of relationship between two variables.

orthogonal—*adjective*. Having no bearing on the question at hand. (Literally, "perpendicular.")

Let's hire these consultants. They have amazing hair.
I'm afraid that's orthogonal to the question of whether they know what they're talking about.

Data

Language nerds, bless their hearts, love to tell us that "data" is plural. The singular is "datum." Thus, you should not use "data" as a mass noun, like "sugar" or "luggage," as in "This data gives corporations total control over my life." You should use it as a plural noun, like "cups" or "buckets," as in "These data give corporations total control over my life." Anyway, I say this concern misses the data for the datums: the point is that as I write these words in the early 2020s, it is increasingly clear that we will spend the rest of our lives in the Century of Data.

variance—*noun*. Unpredictability; variety.

n—*noun*. The number of people from whom data was gathered; the larger the *n*, the more trustworthy the results (all else being equal).

uniformly distributed—*adjective*. With all values being equally likely.

standard deviations above the mean—*plural noun*. Steps above average: one is pretty good, two is very good, three is extremely good, and four is practically off the charts.

representative—*adjective*. Of a small group: resembling the larger group to which it belongs.

We're not giving out salty licorice for Halloween. People hate it.
How dare you! My siblings and I love it!

noisy—*adjective*. Of an outcome or result: influenced or shaped by random chance.

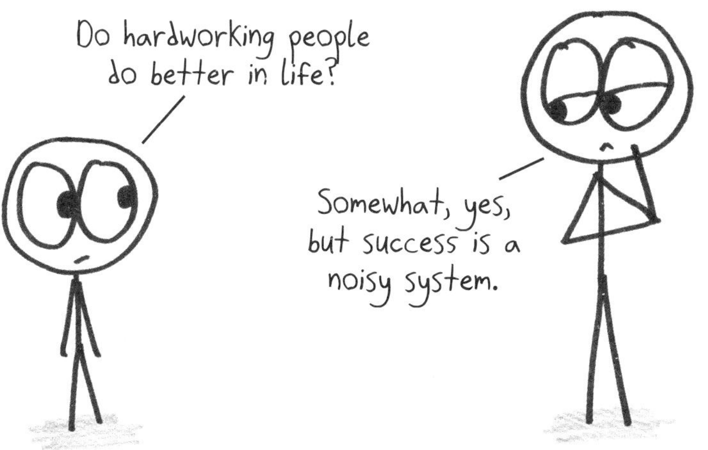

Do hardworking people do better in life?
Somewhat, yes, but success is a noisy system.

null hypothesis—*noun*. The default assumption; something assumed true unless we find compelling evidence to the contrary.

Look at that! I'm telling you, she's a serial killer!
I'm sticking with the null hypothesis that she's a regular person who happens to have a chainsaw in her garage.

Games and Risk

Though it began as a way to dissect simple games of chance, the mathematical field of *game theory* has grown into something far larger: a framework for analyzing any kind of strategic interaction, whether it involves rival athletes, mating lizards, competing corporations, or spacefaring civilizations. As such, the language of game theory won't just help you stop losing at poker night; it'll help you converse intelligently about all manner of topics (while still losing at poker night).

game theory—*noun*. The mathematical study of strategic interactions.

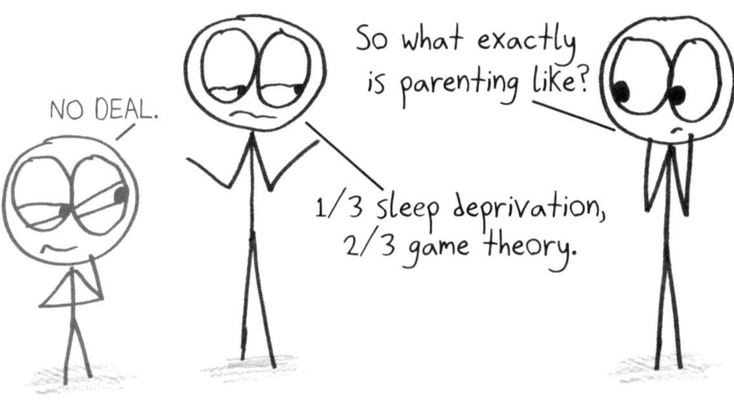

prisoner's dilemma—*noun*. A scenario in which each person faces a choice between the common good and their own self-interest.

zero-sum—*adjective*. A scenario in which one person's gain is necessarily another person's loss.

gambler's fallacy—*noun*. The erroneous belief that if your luck has been bad, good luck will soon come, to "even things out."

expected value—*noun*. The average outcome you'd get if you did something over and over.

risk-averse—*adjective*. Preferring lower risk, even if it means sacrificing expected value.

strictly dominate—*verb*. To be better in at least one way and worse in none.

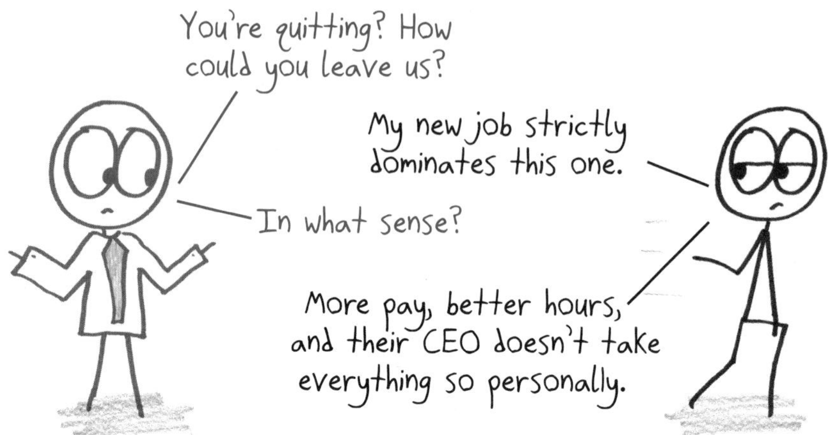

Properties

When someone is too focused on details, we say they're missing the forest for the trees. Mathematicians, if anything, are guilty of the opposite: missing the trees for the forest. Their habit is to inspect not objects themselves, but the objects' properties. (Not the trees, but the number of them.) Then, having distilled these properties, they look at the *properties'* properties. (Not the number, but its evenness or oddness.) And so on: the properties of the properties of the properties of the things. "Matter does not engage their attention," Henri Poincaré once said of mathematicians. "They are interested in form alone."

isomorphic—*adjective*. Sharing the same basic structure, despite apparent differences.

reflexive property—*noun*. The fact that any object is identical to itself.

transitive—*adjective*. Of a connection: if A relates to B, and B relates to C, then A relates to C.

You're my two best friends. Why can't you get along?
Because friendship is not transitive.

invariant—*noun*. A property that doesn't change, even when the object itself does.

You've grown a lot. But the one invariant has been your profound inability to make a bed.

commute—*verb*. To give the same result when the order is switched.

Famous Names and Folklore

Unlike science—which boasts heavy hitters like Marie Curie and Victor Frankenstein—math suffers from a lack of household names. Emmy Noether? Will Hunting? Guest-room names at best. Still, in lieu of public acclaim, mathematicians have something even sweeter: the esteem of their colleagues. Which is to say: inside jokes. If you want to enjoy the silly company of mathematicians, it helps to know a few of these common references.

Erdős number—*noun*. The number of steps removed you are from Paul Erdős, where "one step" means coauthoring a research paper together.

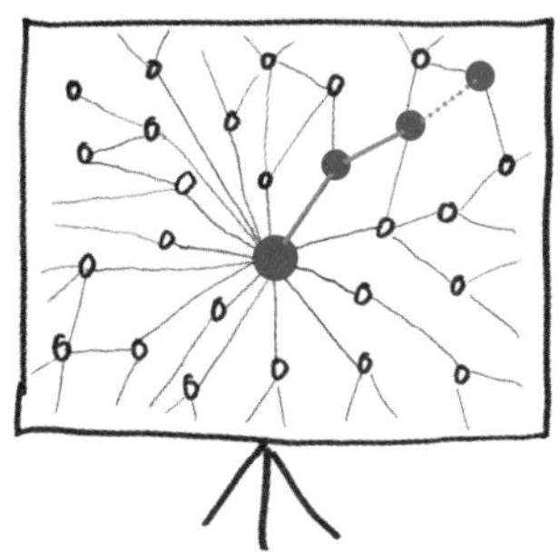

Why are you so eager for us to work together?
Your Erdős number is 2! I've always wanted to be a 3.

Gaussian—*adjective*. Relating to 19th-century mathematician Carl Gauss, who has so many things named after him that you can stick this adjective in front of pretty much any noun.

Ugh, you destroyed me again in fantasy football. Your strategy needs a name.
Oh, it's just a special case of the Gaussian defeat-your-friends theorem.

Fermat's last—*adjective*. Promised but never fulfilled. (After the "last theorem" of Pierre de Fermat, who claimed in the margin of a book to have proved an interesting statement, but with a proof too long to fit in the margin; he was almost certainly mistaken, as it took 350 years after his death to discover a valid proof.)

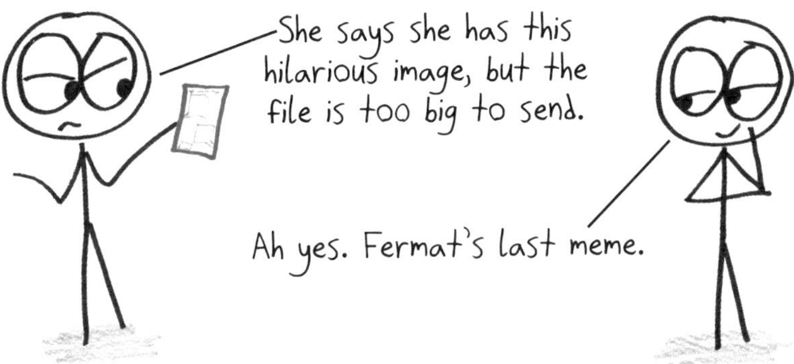

She says she has this hilarious image, but the file is too big to send.
Ah yes. Fermat's last meme.

Fields medal—*noun*. A famous prize in mathematics. Originally intended as an honor for promising up-and-comers. Eventually, they just started giving it to highly accomplished researchers under the age of 40.

I finished my taxes! With no help from software!
Wow! That's Fields-worthy work.
Technically I'm too old, but thank you.

Millennium Problem—*noun*. One of seven famous and important math problems selected in the year 2000. Solving each one brings a $1 million prize.

pull a Perelman—*verb*. To vanish from society. Grigori Perelman solved a Millennium Problem, then refused the prize money and retired from math.

Euler's identity—*noun*. The equation $e^{\pi i} + 1 = 0$. Unifying five fundamental numbers into one simple relationship, it is often cited as the most beautiful equation in all of mathematics.

÷10
10⁻¹ 10⁰ 10¹ 10² 10³
0.1 1 10 100 1,000
π ≈ 3.14
τ ≈ 6.28
y=mx+b
around
across
"857"
-(-(-(-1)))
Ah, simple.
125
+ 124
249
∀ε>0,
∃δ>0:
x−a|<δ→
f(x)−L|<ε
ax²
+++
1/4 + 1/6 =
THE FUZZY TRUTH
96°F
97°F
98.6°F
unreasonable rounding
CHOCODOLLAR
10X10
1/10X10X10
3.14
π
22/7
3.140
3.141
3.142
3.143
farther
closer
17 × whatever
3.4104398
+ 2.5899496
6.0003894
1.414ish
PI DAY PARTY!
3/14
5×3
multiplication
4/7
10
division by 1/2
17
34
division by 2
2573
× 389
3 + −1.78
???
0 5 10 15 20 25 30 35 40 45 50
repeated addition
5⁵⁵
repeated exponentiation
5↑↑3
tetration
NOOOOOO
a ÷ b/c =
4⁷ᐟ² = ???
6
2
whisper conversation blender nightclub thunder
20 30 40 50 60 70 80 90 100 110 120
×1000
x(x+x²)?
x²+x³!
x²=3x+10
8÷2(2+2)
8÷2(4)
8÷8
1
K=½mv²
If A<B and B<C
then A<C
DONUTS
x = −b ± √b² − 4ac / 2a
7 PIZZAS 10/6
44.9° North
93.2° West
ax² = bx + c
ax² + bx = c
ax² + c = bx
ax² = bx
ax² = c
bx = c
1/3 sleep deprivation,
2/3 game theory.
Guaranteed 100%
Almost certain 95% to 99.9%
Very likely 80% to 95%
Probable 60% to 80%
Maybe 40% to 60%
Possible 20% to 40%
Unlikely 5% to 20%
Almost impossible 0.1% to 5%
Impossible 0%
AROUND ≈ 3 × ACROSS
OOOOH!
17/34
noisy
11
"1000"
3
5
1 −1 −(−1) −(−(−1))
5↑↑3 WOW!
30×20 30×7 8×20 8×7
600 + 210 + 160 + 56 = 1026
5+3 5×3 5³
No! Silence your nonsense!
82,771
What is lim x→2 (x²−4)/(x−2)?
0 .1 .2 .3 .4 .5 .6 .7 .8 .9 1
What's 46 + 28?
let's see... calculating... calculating...
27

QUIBBLES, CITATIONS, AND FINE PRINT

Introduction

vii: **Their own utterances:** Oliver Sacks, *Seeing Voices: A Journey into the World of the Deaf* (Berkeley: University of California Press, 1989). Sacks is referring not to mathematical language specifically, but to the nature of being a fluent speaker of any language.

xiii: **"the poetry of logical ideas":** Albert Einstein, "The Late Emmy Noether," *New York Times*, May 4, 1935.

Nouns

2: **"the elaborate cloud language":** Karen Olsson, *The Weil Conjectures: On Math and the Pursuit of the Unknown* (New York: Farrar, Straus and Giroux, 2019).

5: **"Funes the Memorious":** Jorge Luis Borges, *Collected Fictions* (London: Penguin Putnam, 1998).

8: **"to summon a thing that is not there at all":** Ursula Le Guin, *A Wizard of Earthsea* (New York: Houghton Mifflin, 1968).

11: **slips of attention will throw their count off:** For example: Alice Clapman and Ben Goldstein, "Hand-Counting Votes:

A Proven Bad Idea," Brennan Center for Justice, November 23, 2022, https://www.brennancenter.org/our-work/analysis-opinion/hand-counting-votes-proven-bad-idea.

12: **The name "imaginary" originated as a slur:** The mathematician in question was René Descartes. David Wells, *The Penguin Dictionary of Curious and Interesting Numbers* (London: Penguin, 1997).

13: **"I add something to this class":** The student quoted is William Collis. Fittingly enough, his TED talk (with more than 1.5 million views) is "How Video Game Skills Can Get You Ahead in Life." Posted March 2021, https://www.ted.com/talks/william_collis_how_video_game_skills_can_get_you_ahead_in_life.

13: **"A hill can't be a valley":** Lewis Carroll, *Through the Looking-Glass* (1871), Project Gutenberg, https://www.gutenberg.org/files/12/12-h/12-h.htm.

13: **the number line gained currency:** Charalampos Lemonidis and Anastasios Gkolfos, "Number Line in the History and the Education of Mathematics," *Inovacije U Nastavi* 33 (March 2020): 36–56, 10.5937/inovacije2001036L.

14: **Stifel...Bhaskara:** Wells, *Penguin Dictionary of Curious and Interesting Numbers*.

14: **Maseres:** J. J. O'Connor and E. F. Robertson, "Frances Maseres," MacTutor History of Mathematics Archive, accessed May 15, 2023, last updated 2004, https://mathshistory.st-andrews.ac.uk/Biographies/Maseres/.

18: **"Minus times minus equals plus":** W. H. Auden, *A Certain World: A Commonplace Book* (New York: Viking, 1970).

18: **Just ask al-Khwārizmī:** Günhan Caglayan, "Algebra Tiles: Explorations of al-Khwārizmī's Equation Types," *Convergence*, October 2021, https://www.maa.org/press/periodicals/convergence/algebra-tiles-explorations-of-al-khw-rizm-s-equation-types-al-khw-rizm-s-compendium-on-calculating.

26: **Just ask the restaurant chain A&W:** "The Truth About A&W's Third-Pound Burger and the Major Math Mix-Up," A&W Restaurants, accessed April 11, 2023, https://awrestaurants.com/blog/aw-third-pound-burger-fractions. This story gets passed around in math education circles, but it has a whiff of urban legend to it. Maybe the third-pound failed not because the fraction was misunderstood, but because McDonald's is a global behemoth and A&W is known mainly for root beer. In any case, A&W claims this story is true.

32: **the length in words of my book:** Ben Orlin, *Math with Bad Drawings* (New York: Black Dog & Leventhal, 2018). To be honest, I have no idea how many words my book has.

34: **Seife calls this kind of error *disestimation*:** Charles Seife, *Proofiness: The Dark Arts of Mathematical Deception* (New York: Viking, 2010). As selfish writers are wont to do, I've retold Seife's joke in my own words to suit my own rhetorical purposes.

34: **the fiction that the human body temperature is 98.6°F:** Putting together these endnotes, I saw that Carl Wunderlich (the scientist who gave us the 98.6°F figure) actually did specify 37.0°C, to the nearest tenth of a degree. So his number was already too precise, even before its conversion into Fahrenheit.

36: **Frieda's mother, a poet:** That's Claire Wahmanholm, who wrote a lovely essay on the same corn-pit excursion. Claire Wahmanholm, "Get In Loser, We're Going Corn-Pitting," *Essay Daily*, September 20, 2022, http://www.essaydaily.org/2022/09/the-midwessay-claire-wahmanholm-get-in.html. Also deserving a shout-out is Frieda's father, Daniel Lupton, who fact-checked my corn estimate and somehow knows more fun math facts than I do.

37: **The larger roman numerals:** Stephen Chrisomalis, *Reckonings: Numerals, Cognition, and History* (Cambridge, MA: MIT Press, 2020).

39: **The Pacific Ocean holds 10^{20} gallons:** Various semidubious online sources give various semidubious figures, but they're all between 10^{20} and 10^{21}, which is close enough for our purposes. That's typical of my level of precision in this chapter.

39: **A game of chess can unfold in 10^{120} ways:** Claude Shannon, "Programming a Computer for Playing Chess," *Philosophical Magazine*, 7th ser., 41, no. 314 (March 1950).

39: **the *Los Angeles Times* accidentally swapped the words "million" and "billion":** Doug Smith, "But Who's Counting?" *Los Angeles Times*, January 31, 2010, http://articles.latimes.com/2010/jan/31/opinion/la-oe-smith31-2010jan31.

39: **the time a Bear Stearns trader accidentally tried to sell $4 billion in shares:** Floyd Norris, "Erroneous Order for Big Sales Briefly Stirs Up the Big Board," *New York Times*, October 3, 2002, https://www.nytimes.com/2002/10/03/business/erroneous-order-for-big-sales-briefly-stirs-up-the-big-board.html.

39: **the newly elected congressperson who disclosed a family asset worth $1 billion:** Kent Cooper, "Member of Congress Makes Billion-Dollar Error," *Roll Call*, August 20, 2013, https://rollcall.com/2013/08/20/member-of-congress-makes-billion-dollar-error.

39: **"number numbness":** Douglas Hofstadter, *Metamagical Themas: Questing for the Essence of Mind and Pattern* (New York: Basic Books, 1985).

40: **memorizing a vivid image for each number:** John Allen Paulos, *Innumeracy: Mathematical Illiteracy and Its Consequences* (New York: Vintage Books, 1990).

42: **"There are 1,198,500,000 people alive now in China":** Annie Dillard, *For the Time Being* (New York: Vintage Books, 1999).

43: **"There is a size at which dignity begins":** Thomas Hardy, *Two on a Tower: A Romance* (1882). I found the quotation in an excellent pop astronomy book: Marcia Bartusiak, *Dispatches from Planet 3: 32 (Brief) Tales on the Solar System, the Milky Way, and Beyond* (New Haven, CT: Yale University Press, 2018).

45: **"Trapezoid" stems from an error made by some dude in the 1700s:** As Wikipedia explains (https://en.wikipedia.org/wiki/Trapezoid#Etymology_and_trapezium_versus_trapezoid), European languages tend to use variations on "trapezium" for a quadrilateral with one pair of parallel sides, and variations on "trapezoid" for a quadrilateral with no parallel sides. Then, in 1795, Charles Hutton published a mathematical dictionary that accidentally switched the terms. The British corrected the error in the 19th century; the Americans did not.

45: **They're interchangeable:** Some folks insist that "maths" is better because it is plural, reflecting the plurality of mathematical topics (algebra, geometry, etc.). The argument is nonsense, because "maths" is not plural. If it were, we'd say "maths are fun," but we don't; it's "maths is fun." "Maths" (like "mathematics") is a mass noun, which happens to end in an *s*.

47: **a notation for the infinitely small $\frac{1}{\infty}$:** Florian Cajori, *A History of Mathematical Notations* (Mineola, NY: Dover, 1993). Also, while we're talking about notation, I've unilaterally decided to insert a space every three digits after the decimal place, so that instead of writing 3.14159265, we have 3.141 592 65. I don't know why this isn't standard practice already.

49: **from our vantage point, the universe has more small magnitudes than big ones:** The obvious counterargument to this claim is that we can easily imagine bigger magnitudes (say, a distance of 10^{200} light-years or a time span of 10^{500} years), whereas it makes no sense to speak of smaller ones (like 10^{-200} meters or 10^{-500} seconds).

54: **The cool mathematicians all venerate tau:** Vi Hart, "Pi Is (Still) Wrong," YouTube video, uploaded March 14, 2011, https://www.youtube.com/watch?v=jG7vhMMXagQ.

57: **Martin Luther King Jr. having been 5'7"**: This figure is all over the internet, but I honestly have no idea how tall MLK was, nor do I—and this is the key point—care.

59: **"Only three things are infinite":** Gustave Flaubert, *The Letters of Gustave Flaubert, 1830–1857*, trans. Francis Steegmuller (Cambridge, MA: Belknap, 1980).

63: **"Perhaps universal history . . . is the history of a few metaphors":** Jorge Luis Borges, "Pascal's Sphere," in *Other Inquisitions, 1937–1952*, trans. Ruth L. C. Simms (Austin: University of Texas Press, 1964). This is also the source for the Giordano Bruno quote.

64: **"I saw the teeming sea":** Borges, "The Aleph," in *Collected Fictions*.

Verbs

72: **a fourth-tier operation called *tetration*:** You can, of course, push further than tetration. But as you've seen, the numbers just get ridiculously, uselessly, pointlessly large. They really only come up in strange combinatorial settings.

74: **To Orwell's view, simple addition is the last refuge of truth:** George Orwell, *Nineteen Eighty-Four* (London: Secker & Warburg, 1949).

76: **a tale the mathematician Carl Gauss loved to recount:** I cannot recommend more highly the definitive catalogue of retellings of this anecdote: Brian Hayes, "Versions of the Gauss Schoolroom Anecdote," http://bit-player.org/wp-content/extras/gaussfiles/gauss-snippets.html.

91: **How exactly does one multiply?:** Jo Morgan, *A Compendium of Mathematical Methods* (Woodbridge, UK: John Catt, 2019).

97: **"As far as laws of mathematics refer to reality":** Wikiquote attributes this famous line to *Geometrie and Erfahrung* (1921) pp. 3–4, as cited by Karl Popper, *The Two Fundamental Problems of the Theory of Knowledge*, trans. Andreas Pickel, ed. Troels Eggers Hansen (2014).

97: **"Dividing one number by another is mere computation":** Jordan Ellenberg, *How Not to Be Wrong: The Power of Mathematical Thinking* (New York: Penguin, 2014).

106: **we jettisoned the task from the curriculum:** A related task, known as "rationalizing the denominator," has become the subject of one of math education's most inexplicably heated debates. The main justification for this practice (that the long division for $\frac{1}{\sqrt{2}}$ is far nastier than the long division for $\frac{\sqrt{2}}{2}$) is no longer relevant. But a secondary justification (that it's nice to standardize our roots so we can recognize simplifications) still holds. For example, it's baffling that $\sqrt{98} + \sqrt{18} = \sqrt{200}$, but quite natural that $7\sqrt{2} + 3\sqrt{2} = 10\sqrt{2}$.

112: **"mathematics is the art of giving the same name to different things":** Henri Poincaré, *Science and Method* (1908).

113: **documents words from various dialects of English:** David Crystal, *The Disappearing Dictionary: A Treasury of Lost English Dialect Words* (London: Macmillan, 2015).

115: **"A Scottish baron has appeared on the scene":** James Gleick, *The Information: A History, a Theory, a Flood* (New York: Vintage, 2012). This is also the source for the delightful "cheap as potatoes" line.

117: **a cleverly compiled list of ambiguous news headlines:** "Ambiguous Headlines," Fun with Words, http://www.fun-with-words.com/ambiguous_headlines.html. No idea if these are real or fake, but they're pretty fun.

118: ***More powerful operations take priority*:** I have explicitly avoided the traditional acronyms PEMDAS, BODMAS, and BIDMAS, for the simple reason that I hate them. The explanation I give here—*Execute the operations from most powerful to least powerful, unless I say otherwise by using parentheses*—is, to me, the better one.

121: **"artfully perverse, as if constructed to cause mischief":** Steven Strogatz, "That Vexing Math Equation? Here's an

Addition," *New York Times*, August 5, 2019, https://www.nytimes.com/2019/08/05/science/math-equation-pemdas-bodmas.html.

124: **There are 125 sheep and 5 sheepdogs in a flock. How old is the shepherd?:** Kurt Reusser, "Problem Solving Beyond the Logic of Things: Contextual Effects of Understanding and Solving Word Problems," *Instructional Science* 17, no. 4 (1988): 309–38.

Grammar

129: **the pidgin comes to life as a complete language:** Steven Pinker, *The Language Instinct: How the Mind Creates Language* (New York: William Morrow, 1994).

147: **"an early step back from numbers themselves":** Olsson, *Weil Conjectures*.

152: **students often write something else:** Texas A&M University, "Students' Understanding of the Equal Sign Not Equal, Professor Says" (news release), *ScienceDaily*, August 11, 2010, https://www.sciencedaily.com/releases/2010/08/100810122200.htm.

157: **"People have the mistaken impression that mathematics is just equations":** Kristine Larsen, *Stephen Hawking: A Biography* (Westport, CT: Greenwood Press, 2005).

157: **"Nothing beats a good inequality in trying to understand a problem":** Cédric Villani, trans. Malcolm DeBevoise, *Birth of a Theorem: A Mathematical Adventure* (New York: Farrar, Straus and Giroux, 2015). Villani was actually describing the view of his colleague Elliott Lieb, but he seems to endorse the sentiment.

160: **My friend Michael Pershan once gave this advice:** Michael Pershan, "'Draw a Picture' Is Too Darn Abstract for Kids," *Rational Expressions* (blog), August 21, 2014, http://rationalexpressions.blogspot.com/2014/08/draw-picture-is-too-darn-abstract-for.html.

166: **"It is not how much information there is":** Edward R. Tufte, *The Visual Display of Quantitative Information*, 2nd edition (Cheshire, CT: Graphics Press, 2001).

167: **every strawberry on Earth:** John Scalzi, "The Scalzi Theory of Strawberries," *Whatever* (blog), June 8, 2019, https://whatever.scalzi.com/2019/06/08/the-scalzi-theory-of-strawberries.

169: **the Flesch-Kincaid grade level formula:** I made use of the automatic calculator at the website Good Calculators, https://goodcalculators.com/flesch-kincaid-calculator.

169: **a single sentence that lasts a thousand pages:** Katy Waldman, "Can One Sentence Capture All of Life?" *New Yorker*, September 6, 2019, https://www.newyorker.com/books/page-turner/can-one-sentence-capture-all-of-life.

171: **The formula $V + F = E + 2$:** David Richeson, *Euler's Gem: The Polyhedron Formula and the Birth of Topology* (Princeton, NJ: Princeton University Press, 2008).

175: **"The heart and soul of much mathematics":** Barry Mazur, "When Is One Thing Equal to Some Other Thing?" June 12, 2007, Harvard University, https://people.math.harvard.edu/~mazur/preprints/when_is_one.pdf.

176: **"all algebraic manipulation is psychological":** Paul Lockhart, *Measurement* (Cambridge, MA: Belknap, 2012).

176: **"Make everything as simple as possible" . . . "but no simpler":** Actually, what Einstein said was a bit more complicated and context-dependent than this. Still, by his own logic, I think he'd accept the simpler version.

184: **"the idea of the Suez crisis popping out for a bun":** Douglas Adams, *Dirk Gently's Holistic Detective Agency* (London: Pan Books, 1988). The passage is actually not referring to category errors in general, but to the idea of the character Dirk Gently having a friend.

184: **Lemony Snicket offers the example of a waiter:** Lemony Snicket, *The Reptile Room*, book 2 in the series *A Series of Unfortunate Events* (New York: HarperCollins, 1999).

185: **The term "category mistake":** Ofra Magidor, "Category Mistakes," *Stanford Encyclopedia of Philosophy*, eds. Edward N. Zalta and Uri Nodelman, July 5, 2019, https://plato.stanford.edu/archives/fall2022/entries/category-mistakes.

187: **best phrased by architect (and math lover) Piet Hein:** Piet Hein, "The Road to Wisdom," *Grooks* (New York: Doubleday, 1969).

189: **"a lovely little old rectangular green French silver whittling knife":** Mark Forsyth, *The Elements of Eloquence: Secrets of the Perfect Turn of Phrase* (New York: Berkley, 2013).

191: **a chipper young fellow named Kieran:** The anecdote about Kieran first appeared in an essay of mine titled "The Church of the Right Answer," about how to resist the tendency of bureaucratic schooling to choke off curiosity. Ben Orlin, *Math with Bad Drawings* (blog), February 11, 2015, https://mathwithbaddrawings.com/2015/02/11/the-church-of-the-right-answer.

193: **They conclude that text in the form *a(b + c)* can be replaced with the text *ab + ac*:** I wrote about this error more in my essay "Everything is Linear (Or, the Ballad of the Symbol Pushers)." Ben Orlin, *Math with Bad Drawings* (blog), July 8, 2015, https://mathwithbaddrawings.com/2015/07/08/everything-is-linear-or-the-ballad-of-the-symbol-pushers.

194: **"Mathematics is a game played according to certain simple rules with meaningless marks on paper":** The original title of *Math for English Majors* was *Meaningless Marks*.

194: **"How to Avoid Thinking in Math Class":** This was eventually published as a series of four posts on my blog. Ben Orlin, "How to Avoid Thinking in Math Class," part 1, *Math with Bad Drawings* (blog), January 7, 2015, https://mathwithbaddrawings.com/2015/01/07/how-to-avoid-thinking-in-math-class.

195: **"Operations of thought" . . . "are like cavalry charges in battle":** A. N. Whitehead, *An Introduction to Mathematics* (1911), Project Gutenberg, https://www.gutenberg.org/ebooks/41568.

196: **"how the representation of a thing could seem more alluring than the thing itself":** Olsson, *Weil Conjectures*.

Phrase Book

200: **it drew oohs and aahs from US Supreme Court justices:** Specifically, Chief Justice John Roberts and Justice Antonin Scalia. Robert Barnes, "Supreme Court Justices, Law Professor Play with Words," *Washington Post*, January 12, 2010, https://www.washingtonpost.com/wp-dyn/content/article/2010/01/11/AR2010011103690.html.

211: **Gradient descent:** To a novice, it feels like it should be gradient *ascent*, since we're trying to reach a metaphorical hilltop. Indeed, when you're maximizing something, it *is* ascent. But in math, minimization is much more common than maximization: what we're really seeking is the lowest valley. Hence, *descent*.

225: **"a machine for turning coffee into theorems":** Adriana Salerno, "Coffee Into Theorems," *PhD + Epsilon* (blog), American Mathematical Society, April 28, 2015, https://blogs.ams.org/phdplus/2015/04/28/coffee-into-theorems/.

WHERE TO LEARN MORE

For a lyrical tribute to mathematical language: Olsson, Karen. *The Weil Conjectures: On Math and the Pursuit of the Unknown*. New York: Farrar, Straus and Giroux, 2019. It's no accident that I quoted this book throughout my own.

For a helpful, delightful guide to pencil-and-paper mathematics: Morgan, Jo. *A Compendium of Mathematical Methods*. Woodbridge, UK: John Catt, 2019. Great for teachers, great for students, and great for nerds who just want 13 fascinatingly different ways to multiply.

For my favorite works of mathematical fiction: Borges, Jorge Luis. *Collected Fictions*. London: Penguin Putnam, 1998. These stories are like mathematical proofs: dense, rigorous, and universal. Several of them are particularly beloved by mathematicians:

- "Funes, His Memory" (often translated as "Funes the Memorious")
- "The Aleph"
- "The Approach to Al-Mu'tasim"
- "The Library of Babel"
- "The Garden of Forking Paths"

- "The Secret Miracle"
- "The Book of Sand"
- "Blue Tigers"

For the thoughts of the planet's most celebrated mathematician: Tao, Terence. "On Writing." https://terrytao.wordpress.com/advice-on-writing-papers. No mathematician is a household name, but Terry Tao probably comes closest (especially if your house contains another mathematician). His blog includes some thoughtful and generous reflections on mathematical communication. See in particular the posts:

- "Use Good Notation"
- "Take Advantage of the English Language"
- "Give Appropriate Amounts of Detail"
- "On 'Local' and 'Global' Errors in Mathematical Papers, and How to Detect Them"
- "On 'Compilation Errors' in Mathematical Reading, and How to Resolve Them"

For more on "Growth and Change" (p. 201): The relevant subfield here is calculus. I'm vain enough to recommend my own book, alongside one published almost simultaneously:

- Orlin, Ben. *Change Is the Only Constant: The Wisdom of Calculus in a Madcap World*. New York: Black Dog & Leventhal, 2019.
- Strogatz, Steven. *Infinite Powers: How Calculus Reveals the Secrets of the Universe*. New York: Houghton Mifflin Harcourt, 2019.

For more on "Errors and Estimates" (p. 204): I'm combining a few fields of math here—the uncertainty of probability, the confidence intervals of statistics, the error bounding of differential calculus, and the general slipups and mishaps of all mathematical work.

- Parker, Matt. *Humble Pi: When Math Goes Wrong in the Real World*. New York: Riverhead Books, 2020.
- Paulos, John Allen. *Innumeracy: Mathematical Illiteracy and Its Consequences*. New York: Vintage Books, 1990.
- Seife, Charles. *Proofiness: The Dark Arts of Mathematical Deception*. New York: Viking, 2010.

For more on "Optimization" (p. 207): For an unusual but illuminating entry point, I highly recommend: Bosch, Robert. *Opt Art: From Mathematical Optimization to Visual Design*. Princeton, NJ: Princeton University Press, 2019.

For more on "Solutions and Methods" (p. 211): These themes recur across mathematics. If you're curious about algorithms and heuristics, I recommend:

- Fry, Hannah. *Hello, World: Being Human in the Age of Algorithms*. New York: W. W. Norton, 2019.
- Shane, Janelle. *You Look Like a Thing and I Love You: How Artificial Intelligence Works and Why It's Making the World a Weirder Place*. New York: Little, Brown, 2019.

For more on "Shapes and Curves" (p. 214): So many great options, but I recommend:

- Escher, M. C. Translated by Karin Ford. *Escher on Escher: Exploring the Infinite*. New York: Harry N. Abrams, 1989.
- Parker, Matt. *Things to Make and Do in the Fourth Dimension: A Mathematician's Journey Through Narcissistic Numbers, Optimal Dating Algorithms, At Least Two Kinds of Infinity, and More*. New York: Farrar, Straus and Giroux, 2015.
- Roberts, Siobhan. *King of Infinite Space: Donald Coxeter, the Man Who Saved Geometry*. New York: Walker Books, 2006.

For more on "Infinity" (p. 217):

- For a superdense and supertechnical treatment, there's a book by a novelist: Wallace, David Foster. *Everything and More: A Compact History of Infinity*. New York: W. W. Norton, 2003.
- Or for a breezy and conversational treatment, there's a book by a category theorist: Cheng, Eugenia. *Beyond Infinity: An Expedition to the Outer Limits of Mathematics*. New York: Basic Books, 2018.

For more on "Collections" (p. 221): We're talking about set theory, often taken as the logical underpinning of all of mathematics. My favorite intro: Doxiadis, Apostolos, and Christos Papadimitriou. *Logicomix: An Epic Search for Truth*. New York: Bloomsbury, 2009.

For more on "Logic and Proof" (p. 225): The essence of mathematical work. I recommend:

- Cummings, Jay. *Proof: A Long-Form Mathematics Textbook*. Self-published, 2021.
- Nelsen, Roger B. *Proofs Without Words: Exercises in Visual Thinking*. Originally published by Mathematical Association of America, 1993.
- Ording, Philip. *99 Variations on a Proof*. Princeton, NJ: Princeton University Press, 2021.

For more on "Truths and Contradictions" (p. 229): The borderland between mathematics and philosophy is a great place for confusing yourself eternally. To that end:

- A good starting place: Smullyan, Raymond. *What Is the Name of This Book?: The Riddle of Dracula and Other Logical Puzzles*. New York: Penguin Books, 1990.
- A book assembled, amazingly, by a teenaged author: Alsamraee, Hamza E. *Paradoxes: Guiding Forces in Mathematical Exploration*. Curious Math Publications, 2020.

For more on "The Probable and the Possible" (p. 234): When it comes to probability, I'm most interested in how our certainty-craving minds can be coaxed into accepting a life of fundamental uncertainty. Great books in that vein:

- Galef, Julia. *The Scout Mindset: Why Some People See Things Clearly and Others Don't*. New York: Piatkus, 2021.
- Kahneman, Daniel. *Thinking, Fast and Slow*. New York: Farrar, Straus and Giroux, 2013.
- Mlodinow, Leonard. *The Drunkard's Walk: How Randomness Rules Our Lives*. New York: Vintage, 2009.

For more on "Causes and Correlations" (p. 237): As they say, "correlation is not causation." It's true: correlation does not *cause* causation. But the two are certainly correlated.

- For some fantastic, silly fun: Vigen, Tyler. Spurious Correlations. https://www.tylervigen.com/spurious-correlations.
- A simple yet lovely game (especially for Intro Stats teachers such as myself): Guess the Correlation. https://www.guessthecorrelation.com.

For more on "Data" (p. 240): The literature here is so deep I hesitate to even try summarizing it. But some good starting points are:

- Cairo, Alberto. *How Charts Lie: Getting Smarter about Visual Information*. New York: W. W. Norton, 2019.
- Harford, Tim. *The Data Detective: Ten Easy Rules to Make Sense of Statistics*. New York: Riverhead Books, 2021.
- Orlin, Ben. *Math with Bad Drawings: Illuminating the Ideas That Shape Our Reality*. New York: Black Dog & Leventhal, 2018.

For more on "Games and Risk" (p. 244): These terms come from game theory. I'll plug my own book of games here: Orlin, Ben.

Math Games with Bad Drawings: 75¼ Simple, Challenging, Go-Anywhere Games—And Why They Matter. New York: Black Dog & Leventhal, 2022.

For more on "Famous Names and Folklore" (p. 251):

- For more on Paul Erdős, there's a splendid biography: Hoffman, Paul. *The Man Who Only Loved Numbers: The Story of Paul Erdős and the Search for Mathematical Truth*. New York: Hachette, 1998.
- For more on Fermat's last theorem, go read perhaps the finest mathematical popularization ever written: Singh, Simon. *Fermat's Enigma: The Epic Quest to Solve the World's Greatest Mathematical Problem*. New York: Walker, 1997.
- For more on the Fields Medal, try my interview with historian Michael Barany: Orlin, Ben. "The Forgotten Dream of the Fields Medal." *Math with Bad Drawings* (blog), July 25, 2018. https://mathwithbaddrawings.com/2018/07/25/the-forgotten-dream-of-the-fields-medal.
- For more on the Millennium Problems, you can dip into this book (but, caveat lector, these are tough problems to wrap your head around!): Devlin, Keith. *The Millennium Problems: The Seven Greatest Unsolved Mathematical Puzzles of Our Time*. New York: Basic Books, 2002.
- For more on Grigori Perelman and the problem that he helped to solve: Szpiro, George G. *Poincaré's Prize: The Hundred-Year Quest to Solve One of Math's Greatest Puzzles*. New York: Plume, 2007.
- For more on the Riemann hypothesis and the theory behind it: Derbyshire, John. *Prime Obsession: Bernhard Riemann and the Greatest Unsolved Problem in Mathematics*. New York: Plume, 2004.
- For more on Carl Gauss and Leonhard Euler: take pretty much any university-level math class.

BUMBLING WORDS OF GRATITUDE

Usually this page would be called "Acknowledgments," but for this book, that seemed too chilly a term. An "acknowledgment" is a dignified nod; I need to convey more of a weepy hug, a lingering indebtedness. More than just thanked, you ought to feel *uncomfortably* thanked.

First, thanks to my editor Becky Koh, for her guidance on a book that was uniquely frustrating (for both of us) and uniquely rewarding (for me, at least!). This is the math book I've always wanted to write, and I could never have written it without you.

Thanks to the rest of the BD&L team and our skilled accomplices: Betsy Hulsebosch, Kara Thornton, Katie Benezra and Sara Puppala (your work professionalizing my art was particularly heroic on this one), Elizabeth Johnson, Melanie Gold, Zander Kim, Francesca Begos, and any other names I have inexcusably omitted. I have seen this book as a janky folder of text documents; I cannot wait to see it as the actual book that you all bring to life (and to readers).

Thanks to Dado Derviskadic and Steve Troha, whose judgment is much better than my own, and without whom I would not have this bizarre little fantasy career as a professional illustrator who can't draw.

Thanks to my Intro Statistics students at Saint Paul College, to the gracious folks in the tutoring center, to Enyinda Onunwor and Avani Shah, and to my new pals from AMATYC in Toronto. This book, more than any of my others, is inseparable from my work as a teacher, and you all make that work possible (and fun!).

Thanks to the friends and family who gave feedback and encouragement on this book during its halting journey from *Meaningless Marks on Paper* (February 2021) to *The Cartoon Dictionary of Mathematical Symbols* (July 2021) to *How to Speak Math* (January 2022) to *Ugh, I'm Never Going to Finish This: A Math Book That Cannot Be Written and Will Never Exist* (June 2022) to *Math for English Majors* (February 2023). In particular (and please do not let the absence of a name be construed as absence of gratitude): James Orlin, Michael Pershan, David Klumpp, Karen Carlson, Adam Bildersee, Bay Gaillard, Cash Orlin, Jenna Laib, Lark Palermo, Justin Palermo, Andy Juell, Daniel Gala, Seth Kingman, Karen Olsson, Stephen Chrisomalis, Jo Morgan, Tom Burdett, and James Propp. I offer additional, extravagant, blushing gestures of thanks to Grant Sanderson, who expressed faith in this book at a time when I had quite lost it, and Peggy Orlin and Paul Davis, whose wise and patient early readings helped me remember what this book was for.

A final word of love to Taryn, Casey, and Devyn. All the words I write are just little traces and markings of this beautiful life I'm spending with you.

INDEX

A

B

C

D

E

F

G

H

I

N

O

P

Q

R

S

T

U

V

W

Z

Ah, simple.
CHOCODOLLAR
10X10
1/4 + 1/6 =
y=mx+b
√125
+ 124
249
π ≈ 3.14
1/10X10X10
5X3
multiplication
0 5 10 15 20 25 30
repeated addition
3.4104398
+ 2.5899496
6.0003894
3.14 π 22/7
3.140 3.141 3.142 3.143 3.144
farther
closer
division by 1/2
17 34
division by 2
whisper conversation blender
20 30 40 50 60 70 80 90
×1000
÷10
10⁻¹ 10⁰ 10¹ 10² 10³
0.1 1 10 100 1,000
PI DAY PARTY!
3/14
τ ≈ 6.28
$ 10 $
17 × whatever
x² =
8÷2(2+2)
8÷2(4)
8÷8
1
8÷2(4)
4(4)
16
3 + -1.78
82,771
1.414ish
11
17/34
noisy
6
2
around
across
a ÷ b/c = ac/b
+++
-(-(-(-1)))
ax²
96°F 97°F
"857"
4/7
∀ε>0,
∃δ>0:
|x-a|<δ→
|f(x)-L|<ε
$2.50!
5^5^5
repeated exponentiation
5↑↑3
tetration
???